PLASTIC DESIGN OF FRAMES

2

APPLICATIONS

T0335664

PLASTIC DESIGN OF FRAMES

JACQUES HEYMAN

Reader in Engineering
University of Cambridge

2. APPLICATIONS

CAMBRIDGE
UNIVERSITY PRESS

CAMBRIDGE UNIVERSITY PRESS
Cambridge, New York, Melbourne, Madrid, Cape Town, Singapore,
São Paulo, Delhi, Dubai, Tokyo, Mexico City

Cambridge University Press
The Edinburgh Building, Cambridge CB2 8RU, UK

Published in the United States of America by Cambridge University Press, New York

www.cambridge.org
Information on this title: www.cambridge.org/9780521079846

© Cambridge University Press 1971

First published 1971
First paperback edition 2008

A catalogue record for this publication is available from the British Library

ISBN 978-0-521-07984-6 Hardback
ISBN 978-0-521-73087-7 Paperback

CONTENTS

PREFACE

The first volume of Sir John Baker's *The Steel Skeleton* was published in 1954; volume 2 (*Plastic behaviour and design*), with joint authorship, was published in 1956. The two volumes give an account of the development of steelwork design, the first covering the period up to 1936, when the *Recommendations* of the Steel Structures Research Committee were made, and the second taking the story forward to 1954. The books are thus compounded of the history of structural design, of accounts of the important research results, and of the relevant technical advances in the theory of structures.

As soon as volume 2 of *The Steel Skeleton* had been completed, we wished to construct a more orthodox textbook on plastic theory; in the event, volume 1 (*Fundamentals*) of *Plastic Design of Frames* was not published until 1969, again with joint authorship. The present volume completes the original plan, and covers some topics which were not treated in volume 2 of *The Steel Skeleton*.

The first three chapters deal with the notion of the yield surface in the theory of plasticity; the ideas are developed simply and with reference to the frame rather than the continuum, and are applied to reinforced concrete and masonry in chapter 4. The remaining six chapters of the book return to the problem of the plane steel building frame. In chapter 5 is treated the question of elastic–plastic analysis, and in particular the calculation of deflexions. Chapter 6 deals with shakedown, and chapter 7 with minimum-weight design; in both these topics, some advances have been made since the corresponding chapters in *The Steel Skeleton* were written.

Examples are given at the ends of each of the first seven chapters; as in vol. 1, these are roughly graded from easy to difficult.

Chapter 8 discusses methods of numerical analysis, not from the point of view of the construction of detailed programs, but in a way which exposes the analytical skeleton of plastic methods of design. Finally, chapter 9 discusses some of the problems facing the designer of multi-storey buildings, and chapter 10 uses many of the techniques presented in both volumes for the solution of a practical design problem.

Dr A. C. Palmer has read very carefully a draft of the manuscript, and has been quick to spot loose expositions and hurried logic; any remaining imprecision in the book is entirely the fault of the author. Similarly, Mr B. D. Threlfall, who has taken a great deal of time and trouble with the examples, both in the text and at the ends of the chapters, cannot be held responsible for any remaining numerical errors.

1

THE YIELD SURFACE

1.1 The definition of collapse

Chapter 1 of the previous volume started with a general discussion of the kind of behaviour that might be expected of a simply-supported mild-steel beam, loaded transversely out of the elastic into the plastic range. The idealized load–deflexion curve is reproduced in fig. 1.1. As the transverse load W is increased slowly from zero, the central deflexion δ of the beam at first increases elastically and in proportion, so that OA is a straight line. When the load exceeds the value at A, some plastic deformation occurs, with a corresponding greater increase in deflexion. When the point B is reached, the deflexion δ increases without limit at a constant load, say W_C.

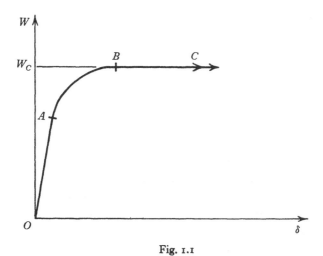

Fig. 1.1

The idealized curve of fig. 1.1 results from the assumption of an elastic–perfectly plastic behaviour of the material; that is, it is assumed that the material does not strain harden as the strain is increased. Strain hardening is in fact small for a structural mild steel, but will always occur eventually; intuitively, the neglect of such increased stresses, at least for relatively small strains, would seem to be 'safe'. It will be assumed throughout the present volume also that the material is elastic–perfectly plastic. (Thus, if the theory is applied practically to a reinforced-concrete structure, for example, then the designer must ensure that the details of the reinforcement are such that any particular plastic hinge does not *strain-soften* due to crushing of the concrete, before deflexions of the structure as a whole become large.)

Similarly, although collapse is characterized by the onset of large deflexions, the assumption will continue to be made that deflexions up to and at incipient collapse are small compared with the overall dimensions of a structure. As was remarked in vol. 1, this assumption implies that

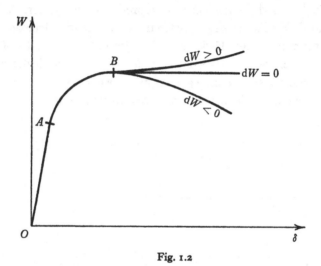

Fig. 1.2

the equations of equilibrium are unchanged after deformation; the shear balance across a storey of a multi-storey frame, for example, is not affected by the small sway of that storey. Thus the deflexions δ in fig. 1.1 are supposed to be very small compared with the span l of the simply-supported beam.

The two essential features of the load–deflexion curve in fig. 1.1 are that a definite collapse load W_C is reached at point B, and that the collapse load stays constant as the deflexions increase from B. The 'flatness' of the collapse portion BC of the curve, along which collapse occurs, may be denoted by $dW = 0$, where dW stands for an infinitesimal variation in the value of the collapse load. By contrast, fig. 1.2 illustrates also the two cases of strain hardening and strain softening; for the former, $dW > 0$, while for strain softening $dW < 0$.

The value of the collapse load of the simply-supported beam may of course be related to the value M_p of the full plastic moment of the beam. The equation of virtual work, using the deflexions marked in fig. 1.3, leads to the relation

$$W\delta = M_p(4\delta/l), \qquad (1.1)$$

which gives directly the familiar expression for the collapse load $W_C = 4M_p/l$. (The deflexion δ in fig. 1.3 is shown as the total central deflexion of the beam from the undeformed state. The corresponding value $4\delta/l$ of the hinge rotation will be correct only if the elastic deflexions of the beam are supposed to be very small compared with the subsequent plastic deflexions. Alternatively, the origin of deflexions might be thought of as at B in fig. 1.1, so that δ measures the plastic deformation only.)

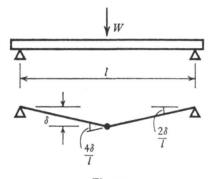

Fig. 1.3

From equation (1.1) it will be seen that the 'ideally plastic' nature of collapse, $dW = 0$, stems directly from the constancy of the value of the full plastic moment M_p. For a given deformation δ of the collapse mechanism, fig. 1.3, all quantities on the right-hand side of equation (1.1) are constant, so that

$$dW.\delta = 0, \qquad (1.2)$$

showing that the small variation dW in the value of the collapse load must be zero.

These ideas are trivial when applied to the case of the simply-supported beam acted upon by a central concentrated load, but have more significant consequences when applied to complex situations. As a first example, the simple rectangular portal frame of fig. 1.4 will be discussed. Here, the frame collapses under the action of *two* loads, V and H, and a yield surface (also called an interaction diagram) was derived in vol. 1 from which the

3

values of the collapse loads can be determined. Figure 1.5 shows such a yield surface for a frame of uniform cross-section, full plastic moment M_p.

It will be recalled that any point in the plane of fig. 1.5 represents a certain combination of loads (V, H). If the point lies within the boundary formed by the three straight lines, then it represents a combination of loads that the frame can support without collapsing. A point on the

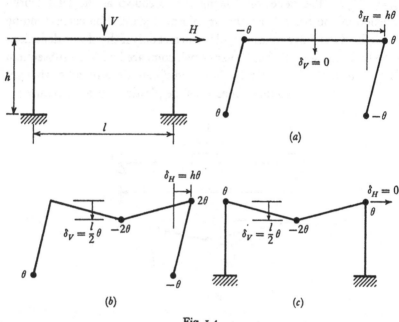

Fig. 1.4

boundary represents a state of collapse, the particular mode depending on the precise ratio V/H for a frame of given dimensions. A point outside the boundary represents loads which cannot be carried by the frame.

Figure 1.5 is incomplete in that it represents only one quadrant of a closed figure round the origin; the complete diagram is, in this case, doubly symmetric, as may be confirmed by investigating collapse mechanisms for negative values of V and H. Suppose attention is confined for the moment to a particular portion of the yield surface, say PQ in fig. 1.5, for which collapse occurs by mode (b) of fig. 1.4. The collapse equation may be written as usual by equating the work done by the loads V and H, during an incremental motion of the collapse mechanism, to the work dissipated in the plastic hinges:

$$V\delta_V + H\delta_H = 6M_p\theta; \tag{1.3}$$

4

the values of δ_V and δ_H may be related to θ as noted in fig. 1.4(b).

For a given motion θ of the collapse mechanism, the values of δ_V and δ_H are fixed, and equation (1.3) leads immediately to the relation

$$dV\,\delta_V + dH\,\delta_H = 0 \qquad (1.4)$$

since the value of M_p is, as before, assumed constant. The ratio dV/dH is the slope of the line PQ, and equation (1.4) may be rearranged as

$$\frac{dV}{dH}\frac{\delta_V}{\delta_H} = -1. \qquad (1.5)$$

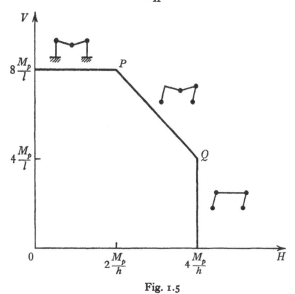

Fig. 1.5

Thus if axes (δ_H, δ_V) are imagined to be superimposed on the axes (H, V) of fig. 1.5, as shown in fig. 1.6, equation (1.5) states that the vector representing the deformation is at right angles to the line PQ.

Equation (1.5) can be checked numerically from the values marked in figs. 1.4(b) and 1.5. From the former, $\delta_V/\delta_H = \tfrac{1}{2}l\theta/h\theta = l/2h$, while from the latter, $dV/dH = (4M_p/l)/(-2M_p/h) = -2h/l$. The *normality condition* illustrated by equation (1.5) and in fig. 1.6 is, as will be seen (equation (1.9)), an essential corollary of the definition of collapse to be made here.

This definition involves first the idea that a definite yield surface, of the type sketched in fig. 1.5, *exists*. That is, if a structure is acted upon by several loads $W_1, W_2, W_3, \ldots$, or, in general, W_j, then there exists a function $f(W_j)$ such that, for $\qquad f(W_j) = \text{const.}, \qquad (1.6)$

5

plastic deformation can occur (of at least part of the structure) which leads to large deflexions. Secondly, the yield surface of equation (1.6) is taken to remain unaltered during collapse, so that the constant on the right-hand side is not affected by the deformations. Under this condition,

$$\Sigma \frac{\partial f}{\partial W_j} dW_j = 0. \qquad (1.7)$$

It will be seen that equation (1.3) is a very simple example of the general equation (1.6), and that equation (1.7) leads immediately to equation (1.4) for this case.

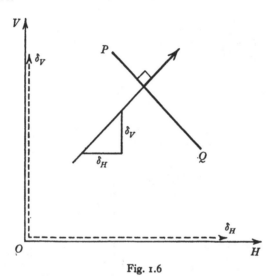

Fig. 1.6

1.2 Characteristics of the yield surface

If collapse of a framed structure occurs according to equation (1.6), then an immediate consequence is that finite deformation of at least part of the structure can occur while the values of the applied loads W_j remain constant. Such deformation will result from the rotation of plastic hinges, and the yield surface $f(W_j)$ can be established by making a plastic work balance for the appropriate collapse mechanism (δ, θ):

$$\Sigma W_j \delta_j = \Sigma M_p |\theta|. \qquad (1.8)$$

The function f is thus piecewise linear (as in fig. 1.5) if all the loads W are concentrated, and the yield surface may be found by identifying the most critical collapse mechanisms for a given frame.

6

As before, since the right-hand side of equation (1.8) is constant for a given motion of a particular collapse mechanism, then

$$\Sigma \mathrm{d}W_j \delta_j = 0. \tag{1.9}$$

Equation (1.9) is alternatively referred to as the principle of maximum plastic work; the loads W_j take values at collapse such that the left-hand side of equation (1.8), written for a given mechanism (δ, θ), is maximized.

Interpreted geometrically, equation (1.9) is a general orthogonality condition between the vectors $\mathrm{d}W$ and δ, which has already been illustrated in two dimensions in fig. 1.6. This relation is the general *normality condition* of plastic theory; although it has been presented here for a perfectly plastic material, the condition holds also if the material strain hardens. Both equations (1.9) and (1.7) must hold for arbitrarily specified values of $\mathrm{d}W_1, \mathrm{d}W_2, \ldots$, so that, comparing the two equations,

$$\delta_j = k \frac{\partial f}{\partial W_j}, \tag{1.10}$$

where k is a factor of proportionality. Thus the normality condition implies that, at collapse, all the displacements δ can be calculated in terms of a single constant k; *ratios* of these displacements can be written at once when the yield surface f, equation (1.6), has been specified.

The discussion so far has been presented in terms of concentrated loads W_j acting upon a framed structure. However, the conclusions are of much wider application. The loads W_j might be interpreted, for example, as the forces acting at a particular cross-section of a beam; a plastic hinge formed under combined bending and twisting will give rise to a yield surface $f(M, T) = \mathrm{const.}$, as will be seen in section 1.4 below. Although this surface is not piecewise linear, the normality condition holds, and the characteristics of the general yield surface, of which fig. 1.5 is a special example, may now be described.

The yield surface is always convex. Thus for a two-dimensional yield curve, fig. 1.7, the straight line joining two points A and B on the curve will lie wholly inside the curve. This geometrical notion of convexity can be extended to the general multi-dimensional yield surface. The proof of convexity perhaps follows most easily from the safe theorem. Briefly, since both points A and B in fig. 1.7 correspond to states of the structure for which it is possible to find equilibrium solutions which do not violate the yield condition (i.e. $|M| \leqslant M_p$ for a frame), then any state found by

7

linear interpolation between these two states will also not violate the yield condition (e.g. if $M^* \leqslant M_p$ and $M^{**} \leqslant M_p$, then

$$[\alpha M^* + (1-\alpha) M^{**}] \leqslant M_p$$

for $0 < \alpha < 1$). Thus point P in fig. 1.7 represents a safe state of loading, and hence by definition is either within or on the yield surface.

These results may be illustrated schematically as in fig. 1.8. The yield surface may contain points or flats, but must nowhere be re-entrant, and it need not be symmetrical. A typical point such as P will define a unique

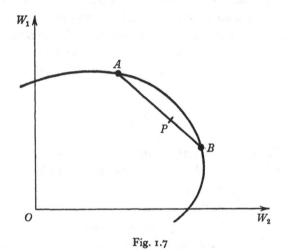

Fig. 1.7

set of loads W_j which will just cause collapse; further, the normal at P will give, to within a factor of proportionality (equation (1.10)), the magnitudes of the corresponding deformations. Along a flat, such as AB, the loading combination is not unique, but the mode of deformation is fixed. For example, in fig. 1.5, the flat PQ corresponds to the mechanism of one degree of freedom sketched in fig. 1.4(b), for which $Hh + \frac{1}{2}Vl = 6M_p$; that is, although the mechanism is fixed, V and H are not determined uniquely.

Similarly, at a point such as C of the yield surface of fig. 1.8, the loads are determined uniquely, but the deformations are not; the 'normal' at the vertex can lie anywhere within the fan shown. Such a vertex corresponds to the simultaneous development of alternative collapse mechanisms for a frame; the point Q in fig. 1.5 corresponds to both mechanisms (a) and (b) of fig. 1.4. These ideas may perhaps be clarified by a numerical

8

example of a simple frame, before applying them to more complex stress situations.

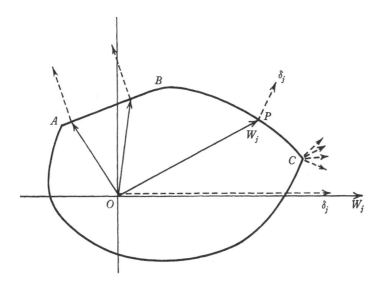

Fig. 1.8

1.3 Frame with distributed load

The uniform frame of fig. 1.9 carries a side load H and a uniformly distributed beam load V. For positive values of V and H there are, as usual, the three possible modes of collapse illustrated. The only difficulty occurs for mode (b), where the position of the hinge within the length of the beam has been specified by the dimensionless parameter z. The general collapse equation for this mode is

$$M_p = \left(\frac{Hh}{2} + z\frac{Vl}{4}\right)\left(\frac{1-z}{2-z}\right). \qquad (1.11)$$

Regarding the problem as one of design for given values of V and H, the value of z must be chosen to maximize the value of M_p, and this condition gives

$$z = 2 - \sqrt{\left[2\left(1 + \frac{Hh}{Vl}\right)\right]}. \qquad (1.12)$$

On substituting this value back into equation (1.11),

$$\left(\frac{Vl}{M_p}\right)^2 + 4\left(\frac{Vl}{M_p}\right)\left(\frac{Hh}{M_p}\right) + 4\left(\frac{Hh}{M_p}\right)^2 - 24\left(\frac{Vl}{M_p}\right) - 16\left(\frac{Hh}{M_p}\right) + 16 = 0. \qquad (1.13)$$

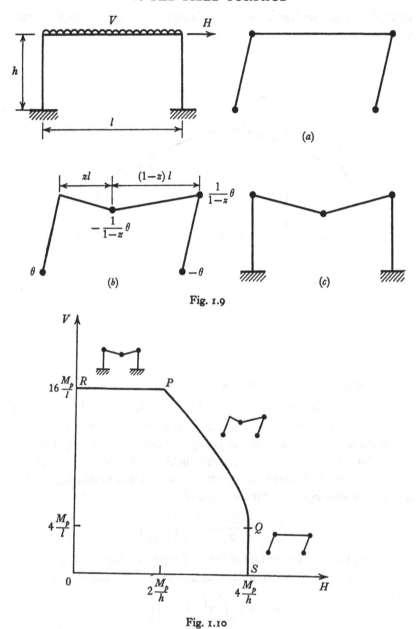

Fig. 1.9

Fig. 1.10

Equation (1.13) is thus the yield surface for this mode, and the complete first quadrant of the yield surface is shown in fig. 1.10.

There are two flats on this yield surface, RP and SQ, and it may be noted that the normal to RP, for example, indicates that $\delta_H = 0$ for this mode. There is no discontinuity of slope at point Q, but there is at P; at the latter point, there is a marked change in the mechanism of collapse, while no such marked change occurs at Q. For any point on the curved portion PQ of the surface, the normal will indicate the deformation pattern; in this case, the horizontal component will give δ_H directly, while the vertical component gives an 'average' δ_V, since the load V is uniformly distributed. As for fig. 1.5, the complete yield surface is doubly symmetric about the origin.

1.4 Combined bending and torsion

Chapter 2 considers the behaviour of a particular type of space frame in which members are acted upon by a single bending moment M and by a twisting moment T. A plastic hinge will form in such members when the values of M and T are such that the relation

$$f(M, T) = \text{const.} \tag{1.14}$$

is satisfied; this expression is, of course, another example of the general relationship (1.6) for the yield surface, and will give rise to an associated flow rule.

The form of the function f in equation (1.14) will first be derived for a thin-walled circular tube of mean diameter d and wall thickness t. If a torque T is applied to this cross-section, the resulting shear stress τ is immediately calculable from equilibrium considerations:

$$T = \tfrac{1}{2}\pi t d^2 \tau. \tag{1.15}$$

If the material of the tube yields according to the Mises criterion, then a longitudinal bending stress σ can be superimposed on the shear stress τ, full plasticity being developed when

$$\sigma^2 + 3\tau^2 = \sigma_0^2, \tag{1.16}$$

where σ_0 is the yield stress in simple tension. Effectively, therefore, the presence of the torque T has reduced the stress available for resisting the bending moment M from σ_0 to σ.

Example 1.1 (c) of vol. 1 gave the plastic modulus of bending of a thin-walled tube as td^2; thus

$$M = td^2\sigma. \tag{1.17}$$

Thus if T and M as given by equations (1.15) and (1.17) just produce a plastic hinge at the cross-section, the corresponding stresses τ and σ are related by equation (1.16), so that

$$\left(\frac{M}{td^2}\right)^2 + 3\left(\frac{T}{\frac{1}{2}\pi td^2}\right)^2 = \sigma_0^2. \tag{1.18}$$

Equation (1.18) may be simplified by introducing the value M_0 of the full plastic moment in the absence of torque, and the value T_0 of the full plastic moment in absence of bending moment:

$$\left.\begin{aligned} M_0 &= td^2\sigma_0, \\ T_0 &= \frac{\pi}{2}td^2\frac{\sigma_0}{\sqrt{3}}. \end{aligned}\right\} \tag{1.19}$$

Equation (1.16) then becomes

$$\left(\frac{M}{M_0}\right)^2 + \left(\frac{T}{T_0}\right)^2 = 1. \tag{1.20}$$

When plotted with axes M/M_0, T/T_0, equation (1.20) is of course a circle of unit radius. Although the equation has been derived for a circular tubular cross-section, it might be expected to hold, perhaps approximately, for many other cross-sections. Indeed for any cross-section for which the bending moment M is linearly and continuously related to a stress σ (as in equation (1.17)), the torque T is similarly related to a shear stress τ (as in equation (1.15)), and where the stresses σ and τ are themselves related by some quadratic relationship

$$\sigma^2 + k\tau^2 = \sigma_0^2, \tag{1.21}$$

expressing the condition for full plasticity, then precisely equation (1.20) will always result. (The constant k in equation (1.21) is equal to 3 if the Mises criterion is used, equation (1.16), and equal to 4 for the Tresca criterion; the first criterion predicts that the yield stress in pure shear is $\sigma_0/\sqrt{3}$, and the second that it is $\frac{1}{2}\sigma_0$.)

More generally, there are definite limitations on the form of a possible yield criterion of the type of equation (1.14), which become more apparent when the equation is rewritten in the form

$$f\left(\frac{M}{M_0}, \frac{T}{T_0}\right) = \text{const.} \tag{1.22}$$

If the yield surface is to be doubly symmetric, for example, which is very often the case in practice, then the range in which equation (1.22) can lie is severely restricted by the convexity condition.

In fig. 1.11 the inner square connecting the unit points on the axes represents the smallest possible boundary which is not re-entrant; similarly, the outer square represents the outer limit of all possible doubly symmetric yield surfaces. It will be seen that equation (1.20) lies

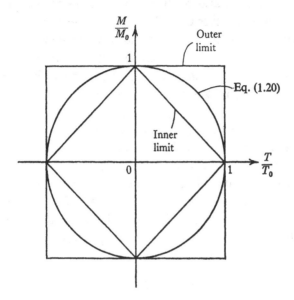

Fig. 1.11

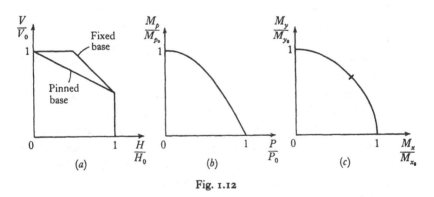

Fig. 1.12

between these two limits, but these remarks hold for the development of full plasticity under any two loading parameters, not necessarily bending moment and torque. For example, the modes of collapse of the fixed-base frame, fig. 1.5, have been replotted in fig. 1.12(a), using axes V/V_0, H/H_0, where, for example, the load H_0 is that necessary to produce sidesway collapse in the absence of vertical load V. Also shown in fig. 1.12(a) is the corresponding yield surface for the pin-based frame (fig. 2.37 of vol. 1); the value of H_0 is not, of course, the same for the two frames.

Similarly, the yield surface of fig. 1.10, for the frame with a uniformly distributed beam load, also lies between the limits marked in fig. 1.11. Fig. 1.12 shows some other cases dealt with in the previous volume. In fig. 1.12(b) is sketched the effect of axial load on a rectangular section bent about a principal axis (fig. 1.27, vol. 1). Figure 1.12(c) is a replot of fig. 1.42, vol. 1, for the bending of a rectangular section about an inclined axis; this particular problem will be discussed more fully in chapter 3 below.

Figures 1.11 and 1.12 indicate that it is fairly easy to construct an approximate two-dimensional yield surface from the results of very few experiments, provided that it can be assumed that the surface is doubly symmetric about the axes. The surface is compelled to pass through the unit points on the axes, and some information is desirable about the slope of the surface as it cuts the axes. For example, the yield surface for combined bending and axial load, fig. 1.12(b), is given by

$$\left(\frac{M_p}{M_{p_0}}\right) + \left(\frac{P}{P_0}\right)^2 = 1; \tag{1.23}$$

this curve cuts the M_p/M_{p_0} axis with zero slope, but makes an angle of $\tan^{-1} 2$ (just over 60°) with the P/P_0 axis. If the slopes at the axes can be determined, either from mathematical considerations or by experiment, then a single point on the curve for approximately equal values of the loading parameters will enable the complete yield surface to be sketched with some confidence.

For the particular problem of combined bending and axial load, an outer bound to the yield surface may be established by using the normality condition. For zero axial load, no axial deformation is possible, and the slope of the yield surface as it cuts the M_p/M_{p_0} axis must therefore be zero. As the bending moment becomes small, so the zero-stress axis approaches the edge of the cross-section, fig. 1.13. In the limit, the ratio of axial deformation to rotation is simply $\frac{1}{2}d$, and this establishes that the yield surface must cut the P/P_0 axis with a slope of $\tan^{-1} 2$. Thus, without deriving the general expression (1.23), the slopes of the yield surface are known at the two axes; since the yield surface cannot be re-entrant, an outer bound may be fixed as in fig. 1.14, which is rather better than that of fig. 1.11.

The yield surface need not necessarily be doubly symmetric, as has been mentioned, nor indeed need it exhibit any symmetry. It was noted in vol. 1 that the yield surface for an unsymmetrical cross-section bent about an inclined axis will in fact be skew-symmetrical; again, this topic

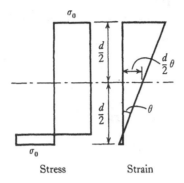

Fig. 1.13

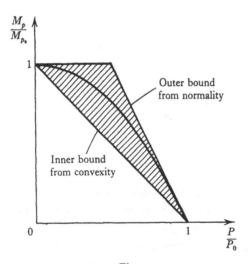

Fig. 1.14

is discussed in chapter 3 later. Another skew-symmetrical surface is obtained for a cross-section having a single axis of symmetry, when bent about an axis at right angles to the axis of symmetry in the presence of axial load; fig. 1.15 reproduces fig. 1.36 of vol. 1.

The yield surfaces, of figs. 1.12 for example, have been sketched for convenience in two dimensions, but the ideas can be extended easily to the discussion of plastic collapse under three or more loading parameters, e.g. to the formation of a plastic hinge under the simultaneous action of axial load and shear force, and possibly of torque. The analogue of the unit circle, equation (1.20), is the unit sphere in three dimensions; the inscribed icosahedron is the minimum possible yield surface if the convexity condition is to be preserved, and all actual yield surfaces must lie outside this minimum surface.

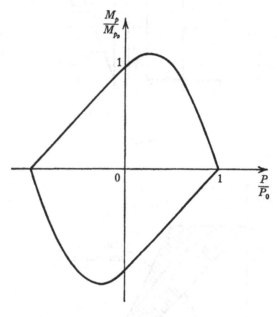

Fig. 1.15

Similarly, the 'interaction diagram' for a frame acted upon by four independent loads may be imagined to be constructed in a four-dimensional loading space. Confining the discussion, for the sake of illustration, to structures whose yield surface is a hypersphere, then the general collapse equation may be written, if there are N loads, as

$$\sum_{i=1}^{N} \left(\frac{W_j}{W_{pj}} \right)^2 = 1. \tag{1.24}$$

The load W_{p_1}, for example, is the value of the load W_1 which would cause collapse of the structure if it acted alone.

In this case the unit hypersphere of equation (1.24) is precisely the function f of equation (1.6). Thus the relative deformations of the structure in the collapse state may be deduced immediately from equation (1.24) by virtue of the flow rule, equation (1.10). Indeed, the motions δ_j of the points of application of the loads W_j are given by

$$\left. \begin{aligned} \delta_1 &= k\frac{\partial f}{\partial W_1} = 2k\frac{1}{(W_{p_1})^2}W_1, \\ \delta_2 &= 2k\frac{1}{(W_{p_2})^2}W_2, \quad \text{etc.,} \end{aligned} \right\} \tag{1.25}$$

where k is a constant (expressing an arbitrary motion of the mechanism of collapse).

Thus, for the very simple case of combined bending and torsion of a circular tube, for which the condition for formation of a plastic hinge is given by equation (1.20), the following relations must hold:

$$\left. \begin{aligned} \theta &= 2k\left(\frac{1}{M_0^2}\right)M, \\ \gamma &= 2k\left(\frac{1}{T_0^2}\right)T, \end{aligned} \right\} \tag{1.26}$$

that is
$$\frac{\gamma}{\theta} = \left(\frac{M_0}{T_0}\right)^2\frac{T}{M}. \tag{1.27}$$

In equations (1.26) and (1.27), the angle θ is the usual bending discontinuity at a plastic hinge, and the angle γ is the torsional discontinuity. Equation (1.27) states that any bending action θ at a hinge formed under combined bending moment and torque must be accompanied by a twisting action γ, and that the ratio γ/θ is specified directly in terms of T and M.

Equation (1.27) is in fact a statement concerning deformations of a structure at collapse, and must be used in order to determine the value of the corresponding collapse load. This is a radical difference from the case of the plane frame collapsing by the formation of plastic hinges according to the one-dimensional yield 'surface' $|M| = M_p$. For the simple plane frame, equilibrium equations alone suffice to give the value of the collapse load; for the case of a multi-dimensional yield surface, the flow rule must usually be used.

The resulting complexity of the calculations will be illustrated in the next chapter by a discussion of elementary space frames. Although a two-dimensional yield surface will be used, the techniques are readily extensible to the general multi-dimensional surface. Some further remarks on yield surfaces will be made in chapter 3.

EXAMPLES

1.1 The rectangular cross-section shown is subjected to an axial compressive load P and to a bending moment M_x about the major axis; the dimensionless stress resultants p and m_x are given by $p = P/P_0$ and $m_x = M_x/M_{x_0}$, where $P_0 = 4ab\sigma_0$ and $M_{x_0} = 2ab^2\sigma_0$. The zero-stress axis lies as shown. Show that, if the cross-section is fully plastic,

$$p = 1 - \beta, \\ m_x = \beta(2 - \beta),$$

so that
$$p^2 + m_x = 1$$

is the equation of the first quadrant of the yield surface (cf. fig. 1.12(b)). Sketch the complete yield surface (fig. 1.27 of vol. 1).

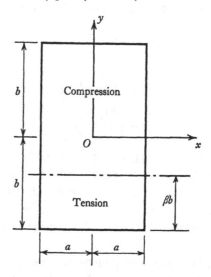

1.2 The symmetrical I-section shown is composed of *thin* rectangles, and is subjected to an axial compressive load P and to a bending moment M_x about the major axis. Show that

$$P_0 = \sigma_0 BT(2 + C),$$

and
$$M_{x_0} = \sigma_0 BTD(1 + \tfrac{1}{4}C),$$

where $C = Dt/BT$.

Show that, for full plasticity with the zero stress axis lying in the two positions indicated, either

$$p = \frac{C(1 - 2\gamma)}{2 + C},$$

$$m_x = \frac{1 + C(\gamma - \gamma^2)}{1 + \tfrac{1}{4}C},$$

so that $m_x + \dfrac{(2 + C)^2}{C(4 + C)}p^2 = 1$,

18

or
$$p = 1 - \frac{2\beta}{2+C},$$
$$m_x = \frac{\beta}{1+\frac{1}{4}C},$$
so that $p + \dfrac{4+C}{2(2+C)}m_x = 1$.

Sketch the first quadrant of the yield surface for $C = 1$.

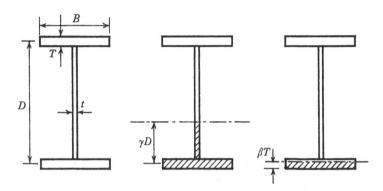

1.3 Repeat example 1.2 for the rectangular hollow section shown; the uniform wall thickness t is small and $C = B/D$.

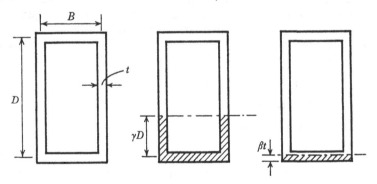

(*Ans.* Either $p = \dfrac{1-2\gamma}{1+C}$,

$m_x = \dfrac{C+2\gamma(1-\gamma)}{\frac{1}{2}+C}$,

$\left.\begin{array}{l}\end{array}\right\}$ $m_x + \dfrac{(1+C)^2}{1+2C}p^2 = 1$,

or $p = 1 - \dfrac{\beta C}{1+C}$,

$m_x = \dfrac{\beta C}{\frac{1}{2}+C}$,

$\left.\begin{array}{l}\end{array}\right\}$ $p + \dfrac{1+2C}{2(1+C)}m_x = 1$.)

2-2

1.4 Repeat example 1.2 for the thin-walled circular tube shown.

$$\left. \begin{array}{l} (\textit{Ans. } p = 1 - \dfrac{2\alpha}{\pi}, \\[2mm] m_x = \sin\alpha, \end{array} \right\} \; m_x = \sin\tfrac{1}{2}\pi(1 - p).$$

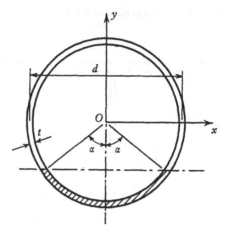

1.5 A thin sheet of steel has a yield stress σ_0 in simple tension. The sheet is subjected to stresses in its plane of σ_x and σ_y at right angles, so that the principal stress system is $(\sigma_x, \sigma_y, 0)$. Using axes σ_x, σ_y, plot the yield surface according to the criteria of (a) von Mises and (b) Tresca. (For a principal stress system $(\sigma_1, \sigma_2, \sigma_3)$ the von Mises criterion is

$$(\sigma_1 - \sigma_2)^2 + (\sigma_2 - \sigma_3)^2 + (\sigma_3 - \sigma_1)^2 = 2\sigma_0^2,$$

and the Tresca criterion is

$$\max\left(|\sigma_1 - \sigma_2|, \; |\sigma_2 - \sigma_3|, \; |\sigma_3 - \sigma_1|\right) = \sigma_0.)$$

(*Ans.* Ellipse, inscribed hexagon.)

1.6 A uniform plate has thickness $2h$ and yield stress σ_0, so that the full plastic moment per unit width of plate is $M_0 = h^2\sigma_0$, and the maximum pull per unit width of plate is $P_0 = 2h\sigma_0$. For the purpose of analysis this uniform plate is replaced by a 'sandwich' plate with the same maximum resistances M_0 and P_0. The sandwich plate consists of two *thin* sheets each of thickness t and yield stress σ_0' separated by a core of thickness $2h'$; the core maintains the separation of the sheets but has no tensile strength. Show that $t\sigma_0' = h\sigma_0$, and $h' = \tfrac{1}{2}h$.

1.7 Using the results of example 1.1, the uniform plate of example 1.6 has a yield surface given by the two curves $p^2 + m = 1$ and $p^2 - m = 1$, when subjected to a bending moment M and an inplane force P. Show that the corresponding yield surface for the equivalent sandwich plate is given by $p \pm m = \pm 1$.

1.8 A sandwich plate (of the type described in example 1.6) is subjected to bending moments M_x and M_y per unit length about axes at right angles. Show that, according to the Mises criterion, the yield surface is the ellipse

$$m_x^2 - m_x m_y + m_y^2 = 1.$$

If the Tresca criterion is used, show that the yield surface is the hexagon

$$m_x = \pm 1, \quad m_y = \pm 1, \quad m_x - m_y = \pm 1.$$

1.9 The sandwich plate is subjected to the dimensionless stress resultants (m_x, p_x) and (m_y, p_y). The general stress distributions in the two directions at right angles are as shown. Show that

$$P_x = t(\sigma_x^+ + \sigma_x^-), \quad M_x = h't(\sigma_x^+ - \sigma_x^-),$$

$$P_y = t(\sigma_y^+ + \sigma_y^-), \quad M_y = h't(\sigma_y^+ - \sigma_y^-),$$

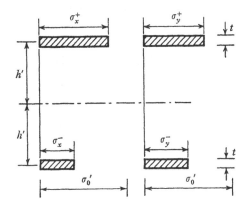

and that, therefore,

$$\sigma_x^+ = \sigma_0'(p_x + m_x), \quad \sigma_y^+ = \sigma_0'(p_y + m_y),$$

$$\sigma_x^- = \sigma_0'(p_x - m_x), \quad \sigma_y^- = \sigma_0'(p_y - m_y).$$

If the Mises criterion is used, show that the yield surface is bounded by the two hypersurfaces

$$(p_x + m_x)^2 - (p_x + m_x)(p_y + m_y) + (p_y + m_y)^2 = 1,$$

$$(p_x - m_x)^2 - (p_x - m_x)(p_y - m_y) + (p_y - m_y)^2 = 1.$$

If the Tresca criterion is used, show that the yield surface is bounded by the twelve hyperplanes

$$p_x \pm m_x = \pm 1,$$

$$p_y \pm m_y = \pm 1,$$

$$(p_x - p_y) \pm (m_x - m_y) = \pm 1.$$

1.10 The web of an I-section is subjected to a bending moment M, an axial thrust P and a shear force S. Assuming that the shear stress due to S is constant, show that the yield surface for the web is given by the equation

$$m^2(1-s^2) = (1-s^2-p^2)^2.$$

(Figure 1.40 of vol. 1 may be found helpful.)

1.11 The frame shown has fixed feet and uniform section, full plastic moment M_p. Using axes (x,y,z), sketch the yield surface (interaction diagram) for the positive octant, and show that it is bounded by the six planes $x = 4$, $y = 4$, $z = 6$, $x+z = 8$, $y+z = 9$, $x+y+z = 11$. Construct the bending moment distribution for the loading $(2, 3, 6)$, and show that this distribution satisfies the yield condition. Determine the value of the collapse load factor for the loading $(2, 2, 3)$. (*Ans.* 1.57.)

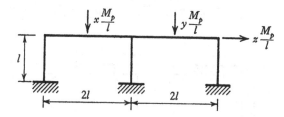

2

ELEMENTARY SPACE FRAMES

The type of frame to be considered in this chapter is in fact a plane frame, but loaded transversely. Thus it will be assumed that all members of the frame lie initially in a certain plane (for convenience, the horizontal plane), and that all loads act in a direction perpendicular to this plane (vertically). Bending moments about vertical axes will be taken to be zero, so that shear forces in the plane are also zero. Any member of the frame will then be acted upon by vertical shear forces, and by two moments about horizontal axes, that is, by a bending moment M (flexural couple) and a torque T (twisting couple).

It will be assumed further (as for the case of the simple plane frame) that the presence of shear forces has no effect on the formation of plastic hinges, so that the yield condition $f(M, T) =$ const. of equation (1.14) may be used. (Shear force may have to be taken into account in any practical design, but the adjustment is relatively easy to make.)

2.1 The right-angle bent

The problem shown in fig. 2.1 was solved in vol. 1 (chapter 5, fig. 5.1), but without any real explanation of how the actual values of bending moment and torque at the hinges were determined. The right-angle bent lies in a horizontal plane, is continuous at B, and is built into fixed walls

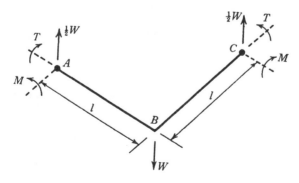

Fig. 2.1

at A and C. The plastic hinges at A and C will form under both bending moment M and torque T, and elementary statics leads to the equation

$$\tfrac{1}{2}Wl = M + T. \tag{2.1}$$

If the members are of uniform tubular cross-section, then the yield condition of equation (1.20) may be written for the hinges A and C:

$$\left(\frac{M}{M_0}\right)^2 + \left(\frac{T}{T_0}\right)^2 = 1. \tag{2.2}$$

Equations (2.1) and (2.2) do not provide enough information for the solution of the problem; that is, M and T cannot be eliminated to give the value W_C of the collapse load.

As mentioned at the end of the previous chapter, the third necessary equation must be found by use of the flow rule, equation (1.27), and this in turn requires consideration of the geometry of deformation of the collapse mechanism. Figure 2.2 is a side elevation of the bent, looking in the original direction BC. If the *bending* rotation at A is denoted by θ, then

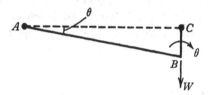

Fig. 2.2

it will be seen that continuity at B demands a *twisting* rotation θ at the hinge at C. Similarly, the bending discontinuity θ at C will be accompanied by a twisting discontinuity θ at A. Thus at A, the ratio of twisting to bending discontinuity is unity, so that equation (1.27) gives

$$1 = \frac{M_0^2}{T_0^2}\frac{T}{M}. \tag{2.3}$$

Equations (2.2) and (2.3) may now be solved simultaneously to give

$$M = \frac{M_0^2}{(M_0^2 + T_0^2)^{\frac{1}{2}}}, \quad T = \frac{T_0^2}{(M_0^2 + T_0^2)^{\frac{1}{2}}}, \tag{2.4}$$

and substitution into equation (2.1) leads to the final expression for the collapse load:

$$\tfrac{1}{2}W_C l = (M_0^2 + T_0^2)^{\frac{1}{2}}. \tag{2.5}$$

(This same value of collapse load was found in vol. 1 by appealing to the principle of maximum plastic work; that is, the value of W given by equation (2.1) was maximized subject to the restriction of equation (2.2).)

The problem shown in fig. 2.3 is very much more difficult, despite its superficial resemblance to the one just solved. For ease of calculation the truly circular criterion

$$M^2 + T^2 = M_0^2 \qquad (2.6)$$

will be used; that is, it will be assumed that $M_0 = T_0$. (This numerical simplification does not restrict the validity of the arguments.) The flow rule associated with equation (2.6) is given from equation (1.27) as

$$\frac{\gamma}{\theta} = \frac{T}{M}. \qquad (2.7)$$

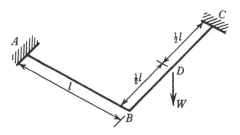

Fig. 2.3

The first trial mechanism will be that of fig. 2.1, as before, with hinges at A and C. The previous geometrical arguments apply to show that $\gamma = \theta$ at each of the hinges, so that, from equation (2.7), $T = M$ at each hinge. Equation (2.6) then gives

$$T = M = \frac{1}{\sqrt{2}} M_0. \qquad (2.8)$$

The work equation for the collapse mechanism of fig. 2.1 (with the load W acting at D, the midpoint of BC), gives

$$W\left(\frac{l}{2}\theta\right) = 2(T\theta + M\theta),$$

that is

$$W = \frac{4\sqrt{2}}{l} M_0. \qquad (2.9)$$

At this stage, all that can be said is that the value of W given by equation (2.9) is an upper bound on the correct value W_C of the collapse load. A statical analysis must be made to confirm that the yield condition, equation (2.6), is nowhere violated.

Figure 2.4 shows the bending moments and torques acting at the hinges A and C; it is an easy matter to calculate the vertical reactions at these two points. Under this loading, conditions at the critical sections B and D must be checked, and, of these two, D is found to be the more critical. By taking moments, the sagging moment at D is determined as $\sqrt{2}M_0$, while the torque at D is the same as that at C, namely $(1/\sqrt{2})M_0$. Thus the left-hand side of equation (2.6) may be written

$$M_D^2 + T_D^2 = (\sqrt{2}M_0)^2 + \left(\frac{1}{\sqrt{2}}M_0\right)^2 = [\sqrt{(\tfrac{5}{2})}M_0]^2. \qquad (2.10)$$

The yield condition is violated at D, and hence the trial solution is incorrect.

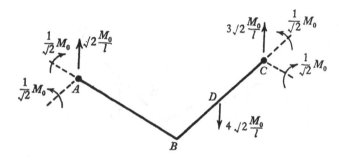

Fig. 2.4

As usual, a lower bound may be found from the result of equation (2.10). If all the numbers in fig. 2.4 are multiplied by $\sqrt{(\tfrac{2}{5})}$, a system of forces, bending moments and torques will be derived which is in equilibrium, and which nowhere violates the yield condition (the yield condition will just be satisfied at D). Thus, from equation (2.9)

$$\frac{8}{\sqrt{5}}\frac{M_0}{l} \leqslant W_C \leqslant 4\sqrt{2}\frac{M_0}{l},$$

that is

$$3 \cdot 58\frac{M_0}{l} \leqslant W_C \leqslant 5 \cdot 66\frac{M_0}{l}. \qquad (2.11)$$

The difficulty now lies in the construction of an alternative collapse mechanism. By analogy with previous work on plane frames, the solution should be improved by the insertion of a sagging hinge at D. One obvious trial would be the simple beam mechanism with hinges at B, C and D, but this leads to the equation $Wl = 8M_0$, which is clearly not the correct solution, from inequalities (2.11). There is no other arrangement of hinges that includes a hinge at D, which gives a mechanism of the usual type of one degree of freedom.

The actual mode of collapse is one which involves hinges at all three points A, C and D. At each of these hinge points two degrees of freedom are permitted; since the original frame had three redundancies, the resulting mechanism, viewed purely as a mechanism, appears to have three degrees of freedom. However, the flow rule must be obeyed at each hinge, and it will be seen immediately that exactly the right number of equations is obtained for the solution of the problem as a whole. Before proceeding with this particular solution, the 'counting' will be done for a general frame of the same type to indicate the genesis of the equations.

At each hinge i of the collapsing frame may be written the two equations:

$$\left.\begin{array}{c} f(M_i, T_i) = \text{const.,} \\[2mm] \dfrac{\gamma_i}{\theta_i} = \dfrac{\partial f/\partial T_i}{\partial f/\partial M_i}; \end{array}\right\} \tag{2.12}$$

the first of these equations is the given yield criterion, and the second is the associated flow rule. If the collapse mechanism has N hinges, then the number of unknown quantities in a general formulation of the problem will be $(4N+1)$; $2N$ of these are unknown bending moments M_i and torques T_i, $2N$ are unknown bending and twisting discontinuities θ_i and γ_i, and the last unknown is the collapse load (or, for the case of proportional loading under many loads, the value of the collapse load factor).

If the number of hinges is such that *all* the hinge discontinuities (bending and twisting) can be written in terms of a single parameter (i.e. if the mechanism is of one degree of freedom in the usual sense), then equations (2.12) may be solved immediately at each hinge to give the values M_i and T_i. A single application of the work equation to the collapse mechanism will then give the value of the collapse load. (This was the procedure exemplified by equations (2.6) to (2.9).) If now *one* extra plastic hinge is inserted into this 'regular' collapse mechanism, *four* extra unknowns (M, T, θ, γ) are introduced into the problem. However, the yield condition and the flow rule, equations (2.12), may be written for the extra hinge, and by taking moments about two axes through the hinge two extra equilibrium equations may be written.

Thus no rule has been derived for the number N of hinges to be expected in a collapse mechanism for a frame of the type considered here. The required number of equations appears to be generated independently of the number of hinges and of redundancies (whereas, for the plane frame, $N = R + 1$, where R is the number of redundancies of the portion of the frame concerned in the collapse). This may be illustrated by continuing with the problem of fig. 2.3.

The mechanism with three hinges, A, D and C, is sketched in fig. 2.5, looking in the original directions BA and BC. The six hinge discontinuities are related by three equations, of which one may be found by matching the deflexion at B:

$$l\theta_A = \frac{l}{2}\theta_C + \frac{l}{2}\gamma_A. \qquad (2.13)$$

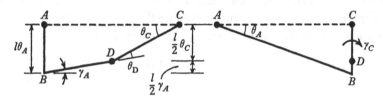

Fig. 2.5

The bending and twisting discontinuities at D are both determinate in terms of the other rotations, so that the three relations may be written:

$$\left.\begin{aligned} 2\theta_A &= \theta_C + \gamma_A, \\ \theta_D &= \theta_C - \gamma_A, \\ \gamma_D &= \theta_A - \gamma_C. \end{aligned}\right\} \qquad (2.14)$$

(Note the 'rigid body' rotation γ_C of the portion CD of the frame.) Since ratios only are required, two unknowns x and y may be introduced, and equations (2.14) rearranged to give

$$\frac{\gamma_A}{\theta_A} = x, \quad \frac{\gamma_C}{\theta_C} = y, \quad \frac{\gamma_D}{\theta_D} = \frac{1 - y(2 - x)}{2(1 - x)}. \qquad (2.15)$$

Equations (2.6) and (2.7), the yield criterion and the flow rule, may now be solved simultaneously for each hinge to give

$$\left.\begin{aligned} M_A &= \frac{M_0}{\sqrt{(1 + x^2)}}, \quad T_A = \frac{xM_0}{\sqrt{(1 + x^2)}}, \\ M_C &= \frac{M_0}{\sqrt{(1 + y^2)}}, \quad T_C = \frac{yM_0}{\sqrt{(1 + y^2)}}, \end{aligned}\right\} \qquad (2.16)$$

with a third pair of more complicated equations for hinge D.

The frame originally had three redundancies, so that a single equation could be written connecting the value of the load W with the four values of bending moment and torque at hinges A and C. The introduction of the extra hinge at D means that two more equilibrium equations can be found which do not involve the value of W; evidently these two equations will serve to determine the values of x and y. One simple equation expresses the fact that the torque in the straight member CD must be constant, so that

$$T_C = T_D,$$

and, therefore, from the yield condition,

$$M_C = M_D. \qquad (2.17)$$

From the flow rule, the ratio γ/θ must be the same at both hinges C and D, so that, from equations (2.15),

$$y = \frac{1 - y(2 - x)}{2(1 - x)},$$

or

$$y(4 - 3x) = 1. \qquad (2.18)$$

Further, equilibrium of portion ABD of the frame requires that

$$T_D + M_A = 2M_D - 2T_A. \qquad (2.19)$$

Introducing the values (2.16), and using (2.17), equation (2.19) leads to a second relation between x and y:

$$\frac{y}{\sqrt{(1 + y^2)}} + \frac{1}{\sqrt{(1 + x^2)}} = \frac{2}{\sqrt{(1 + y^2)}} - \frac{2x}{\sqrt{(1 + x^2)}},$$

or

$$\frac{(1 + 2x)^2}{1 + x^2} = \frac{(2 - y)^2}{1 + y^2}. \qquad (2.20)$$

Finally, equations (2.18) and (2.20) may be solved simultaneously to give the solution

$$x = 0 \cdot 333, \quad y = 0 \cdot 333. \qquad (2.21)$$

(For this simple example, $x = y = \tfrac{1}{3}$, but this identity of x and y is accidental.)

The torques and moments may now be calculated:

$$M_A = M_C = M_D = \frac{3}{\sqrt{10}} M_0 = 0 \cdot 949 M_0; \quad T_A = T_C = T_D = 0 \cdot 316 M_0. \qquad (2.22)$$

Finally, by moments about the line AC, the collapse load W_C is given by

$$W_C l = 2(M_A + T_A + M_C + T_C) = 5 \cdot 06 M_0. \qquad (2.23)$$

The values of the bending moments at collapse, and of the reactions at A and C, are shown in fig. 2.6. The yield condition is satisfied easily at corner B, so that equation (2.23) gives the correct value of the collapse load; note that the value 5.06 lies between the limits of inequalities (2.11).

The solution has been given at some length to demonstrate both that an exact solution can be found for this kind of problem, and also that the working of such a solution may be tedious. For more complex problems it is hardly possible to determine the exact solution, and an approximate

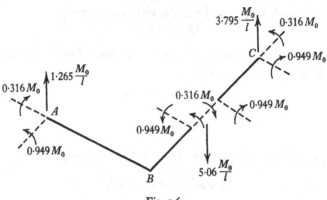

Fig. 2.6

method must be sought. As for the complex plane frame, upper-bound methods relying on assumed mechanisms of collapse are usually quick; they suffer from the disadvantage that they *are* upper bounds, and hence unsafe for design purposes. The simplest way of making the calculations for a given collapse mechanism is to use the work equation, and for this purpose an expression is required for the work dissipated at a plastic hinge.

The general circular criterion at a hinge formed under combined bending and torsion, equation (1.20), may be combined with the associated flow rule, equation (1.27), to give explicit expressions for the bending moment and torque:

$$M = M_0 \left\{ \frac{M_0 \theta}{\sqrt{(M_0^2 \theta^2 + T_0^2 \gamma^2)}} \right\}, \\ T = T_0 \left\{ \frac{T_0 \gamma}{\sqrt{(M_0^2 \theta^2 + T_0^2 \gamma^2)}} \right\}. \tag{2.24}$$

Thus the work dissipated at a plastic hinge, $(M\theta + T\gamma)$, is given by the expression

$$\sqrt{(M_0^2 \theta^2 + T_0^2 \gamma^2)}. \tag{2.25}$$

Equation (2.5), for example, can be deduced immediately by writing expression (2.25) for each of the two hinges, and equating the total to the work done by the load W.

If $M_0 = T_0$, then expression (2.15) simplifies to $M_0\sqrt{(\theta^2 + \gamma^2)}$. Thus, returning to the problem of fig. 2.3, the assumed mechanism of fig. 2.5 may be used in the work equation to give

$$\frac{Wl}{2}\theta_C = M_0\{\sqrt{(\theta_A^2 + \gamma_A^2)} + \sqrt{(\theta_C^2 + \gamma_C^2)} + \sqrt{(\theta_D^2 + \gamma_D^2)}\}. \qquad (2.26)$$

In this equation the six rotations cannot be chosen arbitrarily, since the geometrical relationships (2.14), written alternatively as (2.15), must continue to hold. Subject to these restrictions, however, any values of the rotations inserted into equation (2.26) will lead to a value of W which is an upper bound on the collapse load W_C.

Suppose, for example, that the value of θ_C in fig. 2.5 is taken as unity (to express one degree of freedom of the mechanism), and that it is assumed that BD is horizontal, i.e. that $\theta_D = \theta_C = 1$. The second of equations (2.14) then gives $\gamma_A = 0$ (as is evident from fig. 2.5), and the first gives $\theta_A = 0.5$. If, finally, γ_C is assumed to be zero, then the third of (2.14) gives $\gamma_D = 0.5$, and table 2.1 may be drawn up. From equation (2.26), the value of W is given by

$$W = (3 + \sqrt{5})\frac{M_0}{l} = 5.24\frac{M_0}{l}, \qquad (2.27)$$

which compares with the correct value $W_C = 5.06 M_0/l$.

Table 2.1

Hinge	θ	γ	$\sqrt{(\theta^2 + \gamma^2)}$	M/M_0	T/M_0
A	0.5	0	0.5	1	0
C	1	0	1.0	1	0
D	1	0.5	$\sqrt{(1.25)}$	0.894	0.447

The solution just obtained does not, of course, satisfy equilibrium, and any adjustment which can be made to force the equations to satisfy the statics of the problem is likely to improve the value of the collapse load. Indeed, one way of looking at this approach is to consider an arbitrarily assumed mechanism to be progressively modified until all the equilibrium equations are satisfied. For example, the torques at C and D should be equal, equation (2.17), but they are not so in table 2.1. It is a simple matter to adjust the values of γ_C and γ_D, satisfying the third of equations (2.14) and keeping the other rotations unchanged, in order to ensure that $T_C = T_D$; the results are shown in table 2.2. The corresponding value of W is $(1 + \sqrt{(17)}) M_0/l = 5 \cdot 12 M_0/l$.

Table 2.2

Hinge	θ	γ	$\sqrt{(\theta^2 + \gamma^2)}$	M/M_0	T/M_0
A	0·5	0	0·5	0	0
C	1	0·25	$\sqrt{(1 \cdot 0625)}$	0·971	0·243
D	1	0·25	$\sqrt{(1 \cdot 0625)}$	0·971	0·243

Further modifications may perhaps best be made by trial and error, and an indication of the adjustments to be tried may be found from a consideration of the shear forces corresponding to table 2.2. These are displayed in fig. 2.7, and it will be seen that the solution corresponding

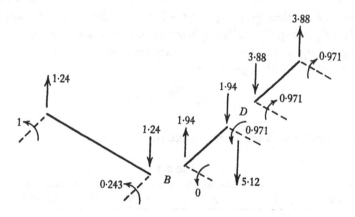

Fig. 2.7

to table 2.2 implies an additional *downward* load of $0 \cdot 70 M_0/l$ acting at the loading point D, while the corner B is evidently subjected to an external *upward* restraining load of the same magnitude. If this upward restraint were removed, it seems plausible that the corner B would drop relative to D, so that the deformation pattern would be more like that sketched in fig. 2.5. Table 2.3 records values of hinge rotations for $\theta_C = 1$, as before, and with γ_C and γ_D chosen so that $T_C = T_D$, but with θ_A increased to the value 0·6. This is, in fact, the correct solution, and gives $W_C = 5 \cdot 06 M_0/l$.

The balancing of shear forces, as in fig. 2.7, will be illustrated further in the discussion below of rectangular grillages.

Table 2.3

Hinge	θ	γ	$\sqrt{(\theta^2+\gamma^2)}$	M/M_0	T/M_0
A	0·6	0·2	0·6325	0·949	0·316
C	1	0·333	1·0541	0·949	0·316
D	0·8	0·267	0·8433	0·949	0·316

2.2 Rectangular grillages

A rectangular grillage of beams, loaded transversely, will collapse in a mechanism involving both bending and twisting of the members, and may be treated by the techniques developed above. It will be assumed again that the circular yield criterion, equation (2.2), together with its associated flow rule, will govern the formation of plastic hinges. Once again, the use of this special criterion (particularly with the further simplification that $M_0 = T_0$) will lead to lighter algebra (and possibly slight changes in precise modes of collapse), while not affecting the general arguments.

In fact, a grillage composed of I-section members could quite well be analysed on the assumption that bending action only is of importance. The full plastic torque T_0 of an I-section is very small compared with the full plastic moment M_0; since torsional rotations γ of any practical collapse mechanism are of the same order as the bending rotations θ, the work dissipated at any hinge, $(M\theta + T\gamma)$, can be approximated fairly accurately by $M_0\theta$. This approximation can be seen to be valid for the circular criterion from expression (2.25); it depends on $(T_0/M_0)^2 \ll 1$. Such an approximate solution was found in chapter 5 of vol. 1 for a simply-supported 3×3 square grillage.

The same square grillage will now be investigated under a different loading system, and the ends of the members will be taken as fixed against both bending and twisting, instead of simply supported. The beams AA, BB, etc., in fig. 2.8 are supposed to be firmly joined where they cross, so that bending moments and torques can be transmitted across such joints. A total vertical load of $9W$ acts on the grillage, split into concentrated loads W at the nine nodes.

If T_0/M_0 is small, the mode of collapse is that shown in fig. 2.9, with hinges formed at the centres of each of the six beams, and no hinges formed at the intersections R. By considering the deflexion of points Q it will be seen that the bending rotation at A must be twice that at B, and elevations of beams AA and BB are shown in fig. 2.10. (It is convenient

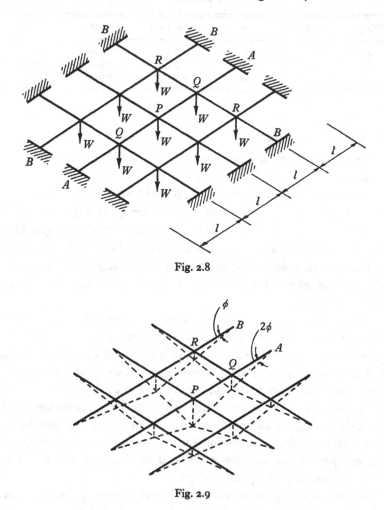

Fig. 2.8

Fig. 2.9

to consider the hinge at the centre of beam AA to be split into two, each with bending rotation 2ϕ, rather than as a single hinge with rotation 4ϕ.) With the displacements shown, the work done by the nine loads W is given by

$$W_{(P)}(4l\phi) + 4W_{(Q)}(2l\phi) + 4W_{(R)}(l\phi) = 16Wl\phi. \qquad (2.28)$$

The table of hinge rotations may be drawn up (table 2.4). Thus, equating

34

the work done by the loads, (2.28), to the sum of the last column of table 2.4, expressing the total work dissipated in the hinges,

$$W = \frac{M_0}{l}\{1 + \sqrt{\{1 + (T_0/M_0)^2\}}\}. \tag{2.29}$$

(It will be seen immediately that, by suppressing the contributions from hinges A and B, the same analysis will give the collapse load of the corresponding simply-supported grillage; the value of the collapse load is exactly half that given by equation (2.29).)

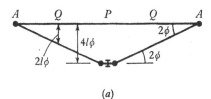

(a)

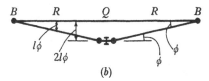

(b)

Fig. 2.10

Table 2.4

Hinge	No.	θ	γ	$\sqrt{(M_0^2\theta^2 + T_0^2\gamma^2)}$
A	4	2ϕ	o	$(4\times)\,2\phi M_0$
P	4	2ϕ	o	$(4\times)\,2\phi M_0$
B	8	ϕ	ϕ	$(8\times)\,\phi\,\sqrt{(M_0^2 + T_0^2)}$
Q	8	ϕ	ϕ	$(8\times)\,\phi\,\sqrt{(M_0^2 + T_0^2)}$

Equation (2.29) gives an upper bound on the true value W_C of the loads at collapse; a statical analysis of the grillage must be made to check whether or not the yield condition is satisfied. Two such analyses will be made, first on the assumption that $T_0 = $ o (i.e. a solution corresponding approximately to an I-beam grillage), and then on the assumption that $T_0 = M_0$ (i.e. the truly circular criterion representative of a box-beam grillage).

(a) $(T_0/M_0)^2 \ll 1$

From equation (2.29), $W = 2M_0/l$. The statical analysis for one-eighth of the grillage is displayed in fig. 2.11, and it will be seen that none of the bending moments exceeds M_0. The bending moment diagram for beams AA is shown in fig. 2.12; that for beams BB is identical. Thus $W_C = 2M_0/l$ is the correct value of the collapse load for the case in which twisting moments can be ignored.

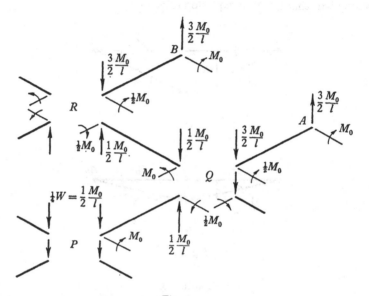

Fig. 2.11

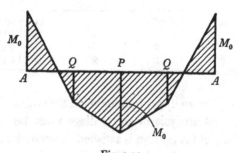

Fig. 2.12

(b) Circular criterion, $T_0 = M_0$

Equation (2.29) gives $W = (1+\sqrt{2})\,M_0/l = 2\cdot414M_0/l$. Using the flow rule, equation (2.7), table 2·4 may be extended as shown in table 2.5. The statical analysis of one-eighth of the grillage is displayed in fig. 2.13; for clarity, the sumbols M_0 and l have been omitted. The yield condition is violated at Q in member QA, the bending moment having value $1\cdot810M_0$. Thus a rather poor lower bound can be established by dividing the results by the factor $1\cdot810$:

$$1\cdot333M_0/l \leqslant W_C \leqslant 2\cdot414M_0/l. \qquad (2.30)$$

Table 2.5

Hinge	θ	γ	$M_0\sqrt{(\theta^2+\gamma^2)}$	M/M_0	T/M_0
A	2ϕ	0	$(4\times)\,2\phi M_0$	1	0
P	2ϕ	0	$(4\times)\,2\phi M_0$	1	0
B	ϕ	ϕ	$(8\times)\sqrt{2}\,\phi M_0$	$1/\sqrt{2}$	$1/\sqrt{2}$
Q	ϕ	ϕ	$(8\times)\sqrt{2}\,\phi M_0$	$1/\sqrt{2}$	$1/\sqrt{2}$

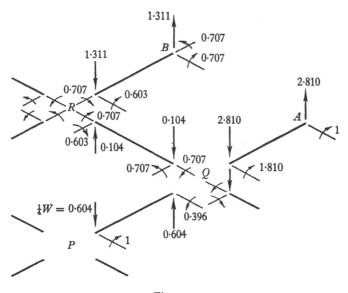

Fig. 2.13

Just as for the right-angle bent, figs. 2.3 and 2.5, the actual collapse mechanism involves the insertion of extra hinges whose rotations have to be determined by use of the flow rule. As before, however, the equations become so complex that a direct solution is difficult, and an assumed deflected shape will be used to give an upper bound on the collapse load.

As a first trial, it will be assumed that the nodes $AQPQA$ lie on a parabola, as do the nodes $BRQRB$, so that the deflected shape of the

37

grillage will be as shown in elevation in fig. 2.14. Thus if the deflexion of node P is, to some arbitrary scale, unity, the deflexions of the nodes will be *assumed* to be

$$\delta_P = 1, \quad \delta_Q = 0.75, \quad \delta_R = 0.5625. \qquad (2.31)$$

With these values, the angles marked in fig. 2.14 may be determined, again to some arbitrary scale, as $\phi_1 = 0.5625$, $\phi_2 = 0.1875$, $\phi_3 = 0.25$. The angle ϕ_4 (giving the twist at hinge B) is not determinate; in table 2.6, the value of ϕ_4 has been chosen to give the same ratio γ/θ at hinges B and D, so that the torque at the two hinges is the same. Thus the two

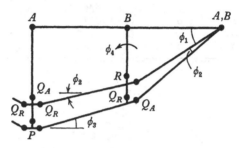

Fig. 2.14

asterisked values in the table do not follow from the assumed deflexions (2.31), but all the other values of θ and γ may be computed in terms of ϕ_1, ϕ_2 and ϕ_3. The total work done in the hinges, $\Sigma M_0 \sqrt{(\theta^2 + \gamma^2)}$, is $15.23 M_0/l$, while the work done by the nine loads W on the deflexions (2.31) is $6.25 W$; equating these two expressions,

$$W = 2.44 M_0/l. \qquad (2.32)$$

Table 2.6

Hinge	No.	θ	γ	$\sqrt{(\theta^2+\gamma^2)}$	M/M_0	T/M_0
A	4	0.75	0	0.75	1	0
Q_A	4	0.5	0	0.5	1	0
P	4	0.25	0	0.25	1	0
B	8	0.5625	0.1125*	0.5737	0.981	0.196
R	8	0.375	0.075*	0.3824	0.981	0.196
Q_R	8	0.1875	0.0625	0.1976	0.949	0.316

This value of W is slightly greater than that given by the simpler mechanism of fig. 2.9, which might perhaps indicate that the true collapse load is fairly close to these upper bounds. A statical analysis of the collapsing grillage in fact indicates the way the mechanism should be modified to improve the value of W.

In fig. 2.15 are marked the bending moments and torques at the hinge points corresponding to the values given in table 2.6, and the statical analysis has been completed for each isolated length of beam forming the grillage. The sums of the shear forces at the nodes are shown boxed; these values should be 2·44 from equation (2.32), that is, they should be equal

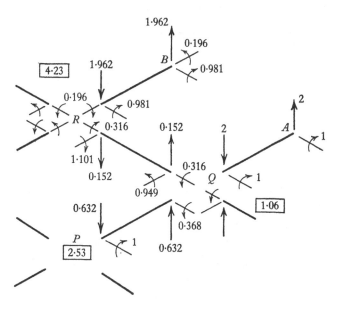

Fig. 2.15

to the load applied at each node. The load at P is about right, but it would seem that node Q has been given too small a deflexion, and node R too large a deflexion. For the next trial, it will be assumed therefore that $\phi_1 = \phi_2$ in fig. 2.14, so that no hinges are formed at R; the deflexions of the nodes will be taken as

$$\delta_P = 1, \quad \delta_Q = 0\cdot 6, \quad \delta_R = 0\cdot 3. \tag{2.33}$$

Corresponding to these deflexions, the values of ϕ may be written as $\phi_1 = \phi_2 = \phi_4 = 0\cdot 3$, $\phi_3 = 0\cdot 4$, and table 2·7 may be constructed.

The total work done in the hinges is $10\cdot 724 M_0/l$, and that done by the loads is $4\cdot 6W$; the corresponding value of W is therefore

$$W = 2\cdot 33 M_0/l. \tag{2.34}$$

39

This is some improvement over the value $2 \cdot 44 M_0/l$ obtained previously. The statical analysis is shown in fig. 2.16; the value of M at node R cannot be found by statics, and has been set equal to $0 \cdot 508 M_0$ in order to give the correct total load ($2.33 M_0/l$) at node R. It will be seen that the total loads carried at P and Q are only slightly different from the correct value.

Table 2.7

Hinge	No.	θ	γ	$\sqrt{(\theta^2+\gamma^2)}$	M/M_0	T/M_0
A	4	0·6	0	0·6	1	0
Q_A	4	0·2	0	0·2	1	0
P	4	0·4	0	0·4	1	0
B	8	0·3	0·3	0·4243	0·707	0·707
Q_R	8	0·3	0·1	0·3162	0·949	0·316

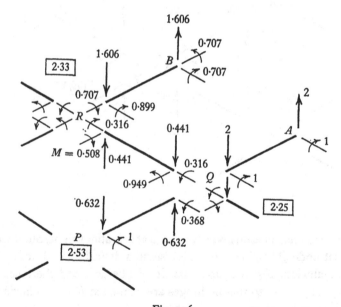

Fig. 2.16

It is not necessary to make a new calculation with a slightly modified mechanism. Small adjustments can be made to the values given in fig. 2.16 so that each node is in equilibrium with the applied load $2.33 M_0/l$. A possible distribution of forces (satisfying equilibrium) is shown in fig. 2.17, where the adjustments have been made in such a way that the yield condition at the hinges B, R and Q_R is violated in the same proportion at each hinge. Thus at B and R, $\{(0 \cdot 811)^2+(0 \cdot 596)^2\}^{\frac{1}{2}} = 1 \cdot 007$, and at Q_R, $\{(0.964)^2+(0.292)^2\}^{\frac{1}{2}} = 1.007$; from equation (2.34) and fig. 2.17, therefore,

$$2.31 M_0/l \leqslant W_C \leqslant 2.33 M_0/l. \tag{2.35}$$

The construction of equilibrium solutions is a tedious process, and the labour becomes very heavy indeed for structures of greater complexity than the 3×3 beam grillage tackled above. However, the derivation of upper bounds remains simple, since hinge discontinuities are readily

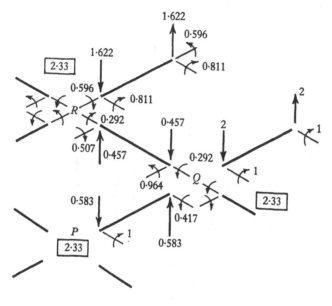

Fig. 2.17

calculable from an assumed deformation pattern, no matter how large and complex the structure. Thus a grillage consisting of 9×9 beams may be analysed quickly and efficiently by assuming the pattern of hinges shown in fig. 2.18; the derivation of this pattern will be clear if comparison is made with the collapse mode of the 3×3 grillage, fig. 2.14. The nodes in fig. 2.18 may be taken as lying on a smooth surface; once deflexions have been assigned to the nodes, both bending and twisting discontinuities are readily calculable.

This way of analysing grillages is analogous to the established procedures for calculating collapse loads of reinforced-concrete slabs by yield-line theory. For the flat slab also it is very difficult to establish equilibrium solutions, and hence to check that a lower bound has been obtained. The work equation, however, which equates the work dissipated in line hinges to the work done by the superimposed loads, enables unsafe designs to be established. The quality and usefulness of such designs depend on the 'reasonableness' of the assumed pattern of yield lines; if the deformation approximates that observed experimentally (perhaps on models) then it is to be expected that reliable estimates of collapse loads can be found. This is an example of one way in which simple experiments can help in design. A pattern of yield lines might be found in the laboratory, and then used as a basis for calculation for the actual structure.

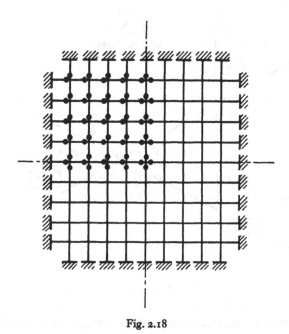

Fig. 2.18

Yield-line theory of reinforced-concrete slabs can itself form the subject matter of whole books,† and will not be dealt with further here. However, the formation of hinges in simple reinforced-concrete beams is of interest in connection with yield surfaces, and will be explored a little in chapter 4.

† For example, K. W. Johansen, *Yield-line theory*, London (Cement and Concrete Association) 1962; R. H. Wood and L. L. Jones, *Yield-line analysis of slabs*, London (Thames and Hudson), 1967.

EXAMPLES

2.1 The frame shown lies in a horizontal plane, and carries a single vertical load W at C. The members are of uniform section with yield criterion $M^2 + T^2 = M_0^2$. $AB = ED = a$; $BC = CD = 2a$. Calculate the value of W to cause collapse. (*Ans.* $\frac{8}{5}(M_0/a)$.)

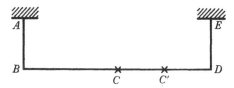

2.2 Repeat example 2.1 for $AB = ED = a$, $BC = CD = ka$.

$$\left(Ans. \ \frac{4k}{k^2+1} \frac{M_0}{a} \text{ for } k \geqslant 1; \ 2\frac{M_0}{a} \text{ for } k \leqslant 1. \right)$$

2.3 The frame of example 2.1 is now loaded transversely with a single load W at the point C', where $BC' = 3a$, $C'D = a$. Find the value of W at collapse.

(*Ans.* $1\cdot79M_0/a$.)

2.4 The horizontal frame shown is composed of three equal members of length $4a$ and yield criterion $M^2 + T^2 = M_0^2$, jointed rigidly together at the centre and built into rigid walls at their other ends. Find the value of the vertical load W, applied as shown, which will just cause collapse. (*Ans.* $1\cdot44M_0/a$.)

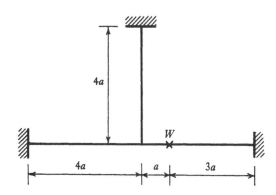

2.5 A square grillage of beams (of the type shown in fig. 2.8) has *four* members in each direction, each of length $5l$. The grillage carries a load W at each node, and the yield criterion of the members is $M^2 + T^2 = M_0^2$. Determine the collapse value of W. (*Ans.* $1\cdot52M_0/l$.)

2.6 Repeat example 2.5 for a 5×5 grillage, each beam having length $6l$.

(*Ans.* $1\cdot12M_0/l$.)

43

2.7 Using the hinge pattern of fig. 2.18, establish bounds for the collapse of a 9×9 grillage. $(Ans.\ 0.40M_0/l \leqslant W \leqslant 0.46M_0/l.)$

2.8 The 3×3 square grillage of fig. 2.8 has members with the yield criterion $M^2 + \alpha T^2 = M_0^2$. Establish the following values of the collapse load W:

α	∞	5	2	1	0
Wl/M_0	2	2·095	2·206	2·322	8/3

3

UNSYMMETRICAL BENDING

3.1 Bending of rectangular section about an inclined axis

The yield surface for a rectangular section bent about an inclined axis was derived in vol. 1. The bending moment M acts about an axis making an angle θ with Ox, fig. 3.1, and is of magnitude such that the whole section is fully plastic. The resulting zero-stress axis does not in general coincide with the axis of the applied bending moment, and it was shown in vol. 1 that the components of the bending moment could be expressed as

$$M_x = M\cos\theta = M_{x_0}(1 - \tfrac{1}{3}z^2), \\ M_y = M\sin\theta = M_{y_0}(\tfrac{2}{3}z), \quad \Bigg\} \tag{3.1}$$

where $M_{x_0}(= 2ab^2\sigma_0)$ and $M_{y_0}(= 2a^2b\sigma_0)$ are the full plastic moments about Ox and Oy respectively; expressions (3.1) hold for $|z| \leqslant 1$, as will be evident from fig. 3.1. Thus, introducing the dimensionless stress resultants m_x and m_y, where $m_x = M_x/M_{x_0}$, and m_y is similarly defined, it will be seen that the yield surface is parabolic, having equation

$$m_x + \tfrac{3}{4}m_y^2 = 1. \tag{3.2}$$

Equation (3.2) holds for $m_x \geqslant \tfrac{2}{3}$, $|m_y| \leqslant \tfrac{2}{3}$, and is plotted (first quadrant) in fig. 1.12(c); a similar equation, with m_x and m_y interchanged, is valid for $|m_x| \leqslant \tfrac{2}{3}$, $m_y \geqslant \tfrac{2}{3}$, that is, for $z \geqslant 1$.

(Some care must be exercised when deducing flow rules from yield surfaces, such as equation (3.2), written in terms of dimensionless stress resultants. Regarding equation (3.2) as the function f defining the yield surface, application of the flow rule, equation (1.10), would give an indicated inclination of the zero-stress axis

$$\tan\alpha = \frac{\partial f/\partial m_y}{\partial f/\partial m_x} = \frac{\tfrac{3}{2}m_y}{1} = z. \tag{3.3}$$

The correct expression, as may be verified from the full equation, is

$$\tan\alpha = z\frac{M_{x_0}}{M_{y_0}} = \frac{b}{a}z, \tag{3.4}$$

which agrees with fig. 3.1; if the real stresses are reduced by certain amounts, then the corresponding deformations must be increased by the same amounts, in order to keep the expression for work done correct.)

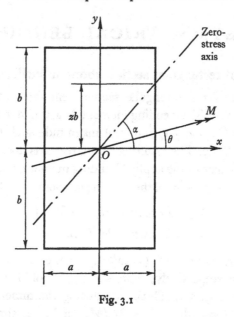

Fig. 3.1

3.2 The effect of axial load

The same rectangular cross-section, fig. 3.1, will now be examined when full plasticity occurs in the presence of an axial load P.† The three-dimensional yield surface will be symmetric about each of the axes M_x, M_y and P, so that only the first octant need be considered, with M_x and M_y positive (i.e. $0 < \theta < \frac{1}{2}\pi$ in fig. 3.1) and with P compressive. Three separate cases must be considered, corresponding to the three distinct positions of the zero-stress axis sketched in fig. 3.2 (hatched portions of the cross-section are yielding in tension).

It is a simple matter to sum the total force across the cross-section to give the value of P, and to take moments about Ox and Oy in turn to give the values of M_x and M_y. For fig. 3.2(a), the results may be written

$$\begin{aligned} p &= 1 - \tfrac{1}{2}(\beta + \gamma), \\ m_x &= (\beta + \gamma) - \tfrac{1}{3}(\beta + \gamma)^2 + \tfrac{1}{3}\beta\gamma, \\ m_y &= \tfrac{1}{3}(\beta - \gamma). \end{aligned} \right\} \tag{3.5}$$

† These results, together with those of examples 3.4, 3.5 and 3.6 at the end of the chapter, were given in a slightly different form by G. A. Morris and S. J. Fenves, Approximate yield surface equations, *Journal of the Engineering Mechanics Division, Proc. A.S.C.E.*, vol. 95, EM4, p. 937, 1969.

The variables β and γ can then be eliminated to give the equation of the yield surface

$$m_x + \tfrac{3}{4}m_y^2 + p^2 = 1. \tag{3.6}$$

The expression is valid for $\beta \leqslant 2$ and $\gamma \geqslant 0$. When the axial load is zero, equation (3.6) reduces at once to equation (3.2); similarly, for bending about the major axis only ($M_y = 0$), the equation reduces to equation (1.23).

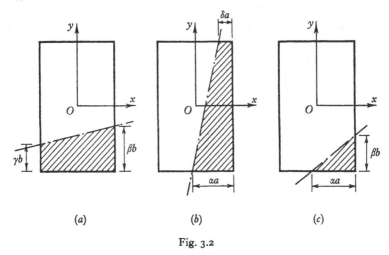

(a)   (b)   (c)

Fig. 3.2

Exactly similar expressions hold for the distribution of fig. 3.2(b), and the corresponding equation for the portion of the yield surface is

$$\tfrac{3}{4}m_x^2 + m_y + p^2 = 1. \tag{3.7}$$

Finally, the distribution of fig. 3.2(c) leads to the equations

$$
\begin{aligned}
p &= 1 - \tfrac{1}{2}(\alpha\beta), \\
m_x &= \tfrac{1}{2}\alpha\beta(1 - \tfrac{1}{3}\beta), \\
m_y &= \tfrac{1}{2}\alpha\beta(1 - \tfrac{1}{3}\alpha),
\end{aligned}
\tag{3.8}
$$

and the corresponding portion of the yield surface is given by

$$\tfrac{6}{5}p + \tfrac{3}{5}p^2 - \tfrac{4}{5}p^3 + \tfrac{9}{10}(1-p)(m_x + m_y) - \tfrac{9}{20}m_x m_y = 1. \tag{3.9}$$

Equations (3.6), (3.7) and (3.9) together define the yield surface in the first octant. Similar surfaces may be calculated for other cross-sections, and examples are given at the end of the chapter.

3.3 The general unsymmetrical section

In fig. 3.3, the axis of the bending moment M makes an angle θ with the direction of Ox of an arbitrary reference frame of axes. If the value of the bending moment is such that the whole cross-section is fully plastic, then, just as for the rectangular cross-section, the zero-stress axis will not in general coincide with the axis of the bending moment. If there is no axial thrust then the zero-stress axis must continue to divide the cross-section into two equal areas. As the axis of the applied bending moment changes the zero-stress axis will shift; it will remain an equal area axis but, for the general unsymmetrical cross-section, it will not pass through a fixed point.

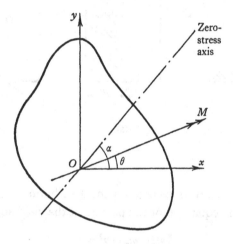

Fig. 3.3

Just as for the rectangular section, the components M_x and M_y of the applied bending moment M may be used as coordinate axes to plot a yield surface (specific calculations are made in section 3.4 for the unequal angle). For the general unsymmetrical cross-section, the yield surface in this coordinate system must be *skew-symmetric*, since the numerical values of M_x and M_y will be the same if the angle θ in fig. 3.3 is increased by π, although the signs will be reversed. Such a skew-symmetric curve is sketched in fig. 3.4. The normal at any point P will make an angle α with the direction of M_x, while OP makes an angle θ. There will be in general two values of θ for which $\alpha = \theta$, that is, for which the axis of deformation is parallel to the axis of applied bending moment; these directions may be called the principal directions of plastic bending, by analogy with the elastic case. The plastic principal axes are indicated in fig. 3.4; they are located by the points of tangency of the inscribed and escribed circles centred on the origin.

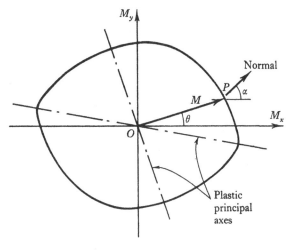

Fig. 3.4

It should be noted that the plastic principal axes do not in general coincide with the corresponding elastic axes; indeed, they are not generally orthogonal. The elastic orthogonality condition follows from the (assumed) linear stress distribution across the section, whatever its shape. In fig. 3.5, the elastic principal axis is shown passing through G, the centre of gravity of the cross-section. If this neutral axis is taken also to be the axis of x, then the bending moment about the axis Gy at right angles must be zero. This requirement leads, on the assumption of a linear stress distribution, to the equation

$$\int^A xy\,dA = 0,\qquad(3.10)$$

where dA is an element of cross-sectional area, and the integration covers the whole cross-section. Equation (3.10) characterizes the elastic principal axes; since the equation is symmetrical in x and y, the directions of these principal axes will be orthogonal.

By contrast, the plastic principal axis sketched in fig. 3.5 does not generally pass through G, but divides the cross-section into two equal areas A_C and A_T, whose centres are C and T respectively. Evidently CGT

is a straight line, and $CG = GT$. If the plastic principal axis is now taken to be the axis of x, then the condition that the axis is indeed *principal* implies, as before, that the bending moment about the y-axis must be zero, that is

$$\int^{A_c} x \, dA = \int^{A_t} x \, dA, \qquad (3.11)$$

where the integrations are carried out over the two separate areas, in tension and in compression.

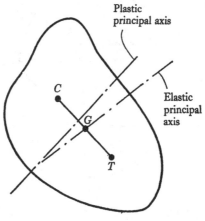

Fig. 3.5

Thus the line CT must be perpendicular to a plastic principal axis, and there are, in general, two solutions of equation (3.11). That is, there will be a 'strong' and a 'weak' principal axis of plastic bending, corresponding to the major and minor axes of elastic bending. However, equation (3.11) is not symmetrical in x and y, so that the principal axes of plastic bending are not compelled to be orthogonal.

3.4   The unequal angle

As an illustration of these ideas, suppose that an unequal angle section is used as a cantilever beam with one leg vertical and the other horizontal, and is loaded at its tip, fig. 3.6. The elastic principal axes are sketched in this figure, and elastic deflexions must be computed by resolving the load into these two directions. On superimposing the resulting deflexions, it is found that the tip of the cantilever moves both horizontally and vertically under the action of a purely vertical load.

Similar behaviour occurs at collapse. The plastic principal axes do not coincide with the corresponding elastic axes, but they are certainly not horizontal and vertical. Thus the cantilever beam of fig. 3.6 will be bent about an axis which is not parallel to a principal plastic axis; at collapse, the tip will again move both horizontally and vertically, although the direction of motion, as will be seen, is not the same at collapse as under an elastic load.

An idealized section, composed of two thin rectangles, will be used for the numerical computation of the plastic principal axes. In fig. 3.7 the

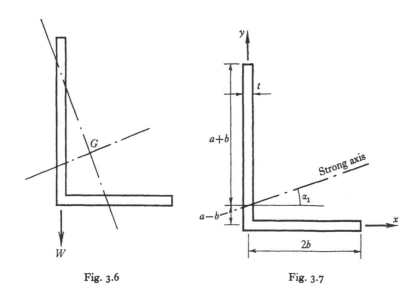

Fig. 3.6   Fig. 3.7

legs have lengths $2a$ and $2b$ and uniform small thickness t; the assumption that t/a (or t/b) is small will introduce slight errors if centre-line dimensions of real angles are substituted into the formulae derived below.

The area of the whole cross-section is $2(a+b)t$; thus the zero-stress axis will divide the cross-section into two equal parts, each of area $(a+b)t$. Figure 3.7 shows the particular case where the zero-stress axis lies in the 'strong' principal direction, making an angle α_1 with the x-axis. Geometrically, the centre of tension T and the centre of compression C could be found for the areas lying on either side of the zero-stress axis (cf. fig. 3.5); the line CT, which is perpendicular to the plastic principal axis, would then give the value of α_1. In fact, it is simpler to compute the bending moments M_x and M_y about the x and y axes, corresponding to the fully plastic state of fig. 3.7, and to combine these to give the required condition.

The following expressions will be found for the values of the bending moments:

$$M_x = (a^2 + 2ab - b^2) t\sigma_0, \\ M_y = 2b^2 t\sigma_0. \quad\Big\}$$

(3.12)

It may be noted that these expressions are independent of the inclination of the zero-stress axis; indeed, from fig. 3.7, the inclination α can lie between $-\tan(a-b)/2b$ and $+\frac{1}{2}\pi$. However, the inclination θ of the axis of the applied moment is fixed from equations (3.12) as

$$\tan\theta = \frac{M_y}{M_x} = \frac{2b^2}{a^2 + 2ab - b^2}.$$

(3.13)

Evidently equations (3.12) correspond to a corner on the yield surface (in fact, they represent point A in fig. 3.9, see later). The value of θ given by equation (3.13) is equal to that of α_1, defining the principal direction.

Using centre-line dimensions, table 3.1 gives some numerical results. The elastic values are 'book' values from the section tables; the inclinations of the principal axes computed for the idealized section of fig. 3.7 are very slightly different from these book values. It will be seen that there is only a small angular difference between the elastic principal axis and the corresponding strong plastic principal axis.

Table 3.1

	a (mm)	b (mm)	$\tan\alpha_1$	α_1	Elastic
Equal angle ($a = b$)	—	—	1	$45°$	$45°$
$6 \times 3 \times \frac{1}{2} \times 21.43$ kg/m	73	35	0·266	$14.9°$	$14.6°$
$4 \times 3 \times \frac{1}{2} \times 16.36$ kg/m	48	35	0·553	$29.0°$	$28.4°$

The 'weak' plastic principal stress axis cuts both legs of the angle. In fig. 3.8 its location is specified by the parameter z, whose value is to be determined; the axis has been drawn as an equal area axis, cutting the section into two halves each of area $(a+b)t$. From the figure, the inclination of the axis is given by

$$\tan\alpha_2 = \frac{z}{a+b-z}.$$

(3.14)

A second expression for $\tan\alpha_2$ may be found by writing the condition that the axis of the applied bending moment is parallel to the zero-stress axis. Corresponding to the position of the zero-stress axis in fig. 3.8, the applied bending moment has components

$$M_x = \{(a^2 - 2ab - b^2) + 2(a+b)z - z^2\}\, t\sigma_0, \\ M_y = (z^2 - 2b^2)\, t\sigma_0. \qquad (3.15)$$

Thus, for the principal direction,

$$\tan\alpha_2 = \frac{(a^2 - 2ab - b^2) + 2(a+b)z - z^2}{2b^2 - z^2}. \qquad (3.16)$$

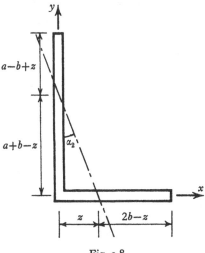

Fig. 3.8

The simultaneous solution of equations (3.14) and (3.16) results in a cubic equation in z having one real root; table 3.2 gives some numerical results. The direction of the weak plastic principal axis is somewhat different from that of the corresponding elastic principal axis for both the unsymmetrical sections. Further, the strong and weak axes are not orthogonal, but intersect at under 80° for both sections. (The results for the 6 × 3 angle will be in error due to the very small inclination of the weak zero-stress axis to the leg; the various portions of the legs in fig. 3.8 have been approximated by rectangles in deriving equations (3.15).)

Table 3.2

	z (mm)	$\tan\alpha_2$	α_2	Elastic
Equal angle ($a = b$)	a	1	45°	45°
6 × 3 × ¼ × 21·43 kg/m	5	0·051	2·9°	14·6°
4 × 3 × ¼ × 16·36 kg/m	20	0·324	18·0°	28·4°

The general case of plastic bending about *any* axis (not a principal plastic axis) can now be discussed with reference to the yield surface for the cross-section. In fig. 3.9, equations (3.15) plot to give the curve AB; the same equations, with signs reversed (bending in the opposite sense) give the curve $A'B'$. At the point B the value of z (fig. 3.8) is zero, and for the portions $A'B$ and AB' of the yield surface the zero-stress axis lies wholly within the longer leg of the angle section. Within the approximation made here (thin rectangles) this means that the value of M_x is indeterminate, and $A'B$ and AB' are straight lines parallel to the axis of M_x. As noted above, equations (3.12) plot as the single point A in fig. 3.9.

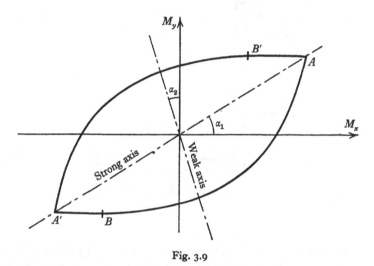

Fig. 3.9

The general shape of fig. 3.9 will be preserved for any angle section; the particular proportions of the figure correspond to a ratio $a/b = \frac{4}{3}$. The strong and weak axes are marked; the latter is, of course, normal to the yield surface. The direction of deformation for bending about the strong axis will depend markedly on very slight variations of the ratio M_y/M_x.

The difference in inclination between the weak plastic axis and the corresponding elastic principal axis, table 3.2, is large enough to be observed experimentally[†]; this difference is over 10° for the 4 × 3 angle. If this angle section is mounted as a cantilever carrying a concentrated tip load, and oriented so that bending occurs about an axis close to the weak axis, then both vertical and lateral movement of the tip will occur in

† J. Heyman, The simple plastic bending of beams, *Proc. Instn civ. Engrs*, 1968, **41**, 751.

general. However, if the angle were mounted so that the value of α in the inset figure in fig. 3.10 were exactly 28·4°, then no lateral movement should occur so long as the material remained elastic. As the full plastic moment was approached the axis of bending would gradually shift towards its final position, given by $\alpha_2 = 18\cdot0°$ in table 3.2, and this shift would be accompanied by lateral deflexion of the tip of the cantilever.

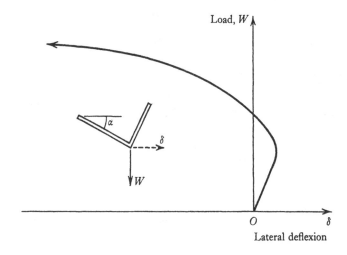

Fig. 3.10

If, on the other hand, the test were arranged so that the value of α in fig. 3·10 lay somewhere between the 18·0° and 28·4° of table 3.2, then the response of the cantilever would be an initial lateral deflexion in one direction, followed by a reversal of this direction as the material became plastic. Figure 3.10 shows schematically the results corresponding to an actual test.

EXAMPLES

3.1 Establish equations (3.1) and (3.2) for the bending about an inclined axis of a rectangular cross-section.

3.2 Equation (3.2), together with the symmetrical expression for $z > 1$, form the yield surface of fig. 1.12(c). For the purpose of analysis, this true yield surface is approximated by the circular criterion $(M_x/M_{x_0})^2 + (M_y/M_{y_0})^2 = 1$. Find the consequent error in the value of the full plastic moment for the point $(M_x/M_{x_0}) = (M_y/M_{y_0})$. *(Ans.* 6 per cent.)

3. UNSYMMETRICAL BENDING

3.3 Establish equations (3.5) and (3.6) for the bending about an inclined axis in the presence of axial load of a rectangular cross-section.

3.4 The I-section shown is composed of *thin* rectangles (cf. example 1.2) and is subjected to an axial compressive load P and to a bending moment about an inclined axis having components M_x and M_y. Five separate cases of full

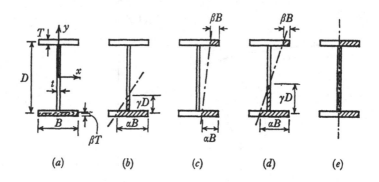

(a)   (b)   (c)   (d)   (e)

plasticity, (a) to (e) in the figure, must be considered to establish the yield surface in the first octant. Prove the following relations ($C = Dt/BT$):

(a)
$$p = 1 - \frac{2\beta}{2+C},$$
$$m_x = \frac{\beta}{1+\frac{1}{4}C},$$
$$\left.\right\} \quad p + \frac{4+C}{2(2+C)} m_x = 1.$$
$$(m_y \text{ indeterminate}),$$

(b)
$$p = 1 - \frac{2(\alpha+\gamma C)}{2+C},$$
$$m_x = \frac{\alpha + C(\gamma - \gamma^2)}{1+\frac{1}{4}C},$$
$$m_y = 2\alpha(1-\alpha).$$

(c)
$$p = 1 - \frac{2(\alpha+\beta)}{2+C},$$
$$m_x = \frac{\alpha-\beta}{1+\frac{1}{4}C},$$
$$m_y = 2\{\alpha(1-\alpha) + \beta(1-\beta)\}.$$

56

(d)
$$p = 1 - \frac{2(\alpha + \beta + \gamma C)}{2 + C},$$

$$m_x = \frac{(\alpha - \beta) + C(\gamma - \gamma^2)}{1 + \frac{1}{4}C}, \qquad \left.\right\} \quad \gamma = \frac{\alpha - \frac{1}{2}}{\alpha - \beta}.$$

$$m_y = 2\{\alpha(1 - \alpha) + \beta(1 - \beta)\},$$

(e)
(p indeterminate),

(m_x indeterminate),

$m_y = 1.$

3.5 Repeat example 3.4 for the rectangular hollow section shown; the uniform thickness t is small, and $C = B/D$.

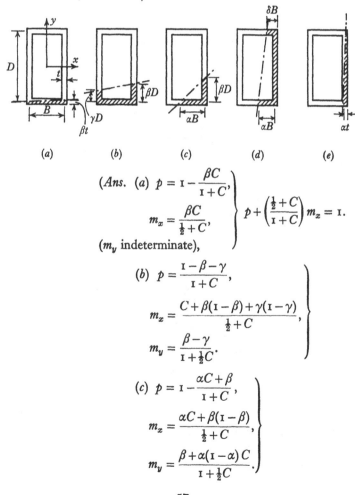

(a)   (b)   (c)   (d)   (e)

(*Ans.* (a) $p = 1 - \dfrac{\beta C}{1 + C}$,

$$m_x = \frac{\beta C}{\frac{1}{2} + C}, \qquad \left.\right\} \quad p + \left(\frac{\frac{1}{2} + C}{1 + C}\right) m_x = 1.$$

(m_y indeterminate),

(b) $p = \dfrac{1 - \beta - \gamma}{1 + C}$,

$$m_x = \frac{C + \beta(1 - \beta) + \gamma(1 - \gamma)}{\frac{1}{2} + C},$$

$$m_y = \frac{\beta - \gamma}{1 + \frac{1}{2}C}.$$

(c) $p = 1 - \dfrac{\alpha C + \beta}{1 + C}$,

$$m_x = \frac{\alpha C + \beta(1 - \beta)}{\frac{1}{2} + C},$$

$$m_y = \frac{\beta + \alpha(1 - \alpha)C}{1 + \frac{1}{2}C}.$$

3. UNSYMMETRICAL BENDING

$$(d)\ \ p = \frac{C(1-\alpha-\delta)}{1+C},$$

$$m_x = \frac{C(\alpha-\delta)}{\frac{1}{2}+C},$$

$$m_y = \frac{1+C\{\alpha(1-\alpha)+\delta(1-\delta)\}}{1+\frac{1}{2}C}.$$

$$(e)\ \ p = 1 - \frac{\alpha}{1+C},$$

$(m_x\ \text{indeterminate}),$ $\quad p+\left(\dfrac{1+\frac{1}{2}C}{1+C}\right)m_y = 1.)$

$$m_y = \frac{\alpha}{1+\frac{1}{2}C},$$

3.6 Repeat example 3.4 for the thin-walled circular tube shown.

$(Ans.\ p = 1 - \dfrac{2\alpha}{\pi},$

$m_x = \sin\alpha\sin\theta,$ $\quad m_x^2 + m_y^2 = \sin^2(\tfrac{1}{2}\pi)(1-p).)$

$m_y = \sin\alpha\cos\theta,$

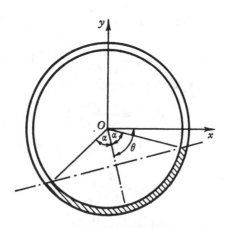

3.7 The Z-section purlin shown is folded from thin uniform sheet. Plot the yield surface in a diagram similar to that of fig. 3.9 for the unequal angle. Determine the angles α_1 and α_2 defining the directions of the plastic principal axes. Calculate the corresponding angle of the elastic principal axes.

$(Ans.\ 18\cdot4°,\ 17\cdot4°,\ 22\tfrac{1}{2}°.)$

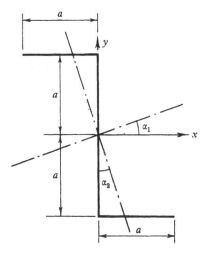

3.8 An angle cantilever was tested as in fig. 3.10 with $\alpha = 26\cdot4°$; the centre-line dimensions of the cross-section were $2a = 93\cdot5$ mm, $2b = 68\cdot5$ mm (fig. 3.7). The resulting deflexions may be plotted as shown (cf. fig. 3.10). Determine the angle β of the plastic axis of bending. (The following steps in the calculation may be found helpful: Calculate the value of z from equations (3.15), using the fact that $M_x/M_y = -\tan 26\cdot4°$. Determine the value of α_2 from equation (3.14), and hence the difference between α_2 and $26\cdot4°$.)

$$(Ans. \; z = 22\cdot2 \, \text{mm}, \; \alpha_2 = 20\cdot7°, \; \beta = 5\cdot7°.)$$

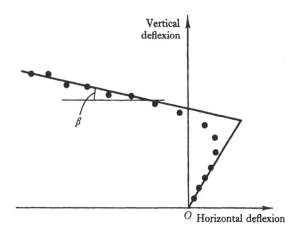

4

REINFORCED CONCRETE
AND MASONRY

If a simply-supported reinforced-concrete beam is tested to collapse, the load–deflexion curve exhibits many of the characteristics associated with the results of a test on a corresponding mild steel beam. Under normal circumstances, for example, the behaviour of the concrete beam is not *brittle*; yielding of the steel reinforcement will permit large deflexions, and a more or less definite plastic hinge may be observed. At very large deflexions the concrete starts to crush, and the resisting moment at the plastic hinge will fall; the length of the 'plateau', corresponding to a notional full plastic moment, and the rate of fall-off of bending moment at large hinge rotations, are functions of the amount and disposition of the reinforcement.

This 'drooping characteristic' of reinforced concrete renders strictly invalid all the theorems of simple plastic theory which are such powerful tools in the analysis of ductile structures. In a *practical* sense, however, the application of simple plastic theory to reinforced-concrete design will lead to perfectly satisfactory structures; or, at worst, it will lead to designs which can be examined carefully in the light of any other criteria the designer may choose to use. As was noticed on the closing page of vol. 1, a plastic estimate of structural behaviour can be used with at least as much confidence as a conventional elastic estimate. An elastic design to a suitable factor of safety may be expected to give a very similar reinforced-concrete structure to a plastic design using a suitable load factor; if the one design is satisfactory, then so will be the other. The only doubt is whether reinforced concrete has sufficient ductility to be a safe structural material, and this is a doubt to be resolved independently of the method used for design. For practical purposes the enormous amount of experience with reinforced-concrete structures leaves no doubt of the safety of the material.

4.1 The simple plastic hinge

The typical stress–strain behaviour of unreinforced concrete in compression is shown, schematically, in fig. 4.1. It will be assumed in all the
following work that concrete cracks in tension, and so is incapable of
carrying any tensile stresses; this is the conventional simple assumption
for elastic design. (It will also be assumed that, at any particular cross-
section of a structure, separate provision is made for any shear forces
that might be present; it is, in fact, possible to make allowance for shear
stresses in the concrete, but it will be taken that direct stresses only act at
a cross-section.)

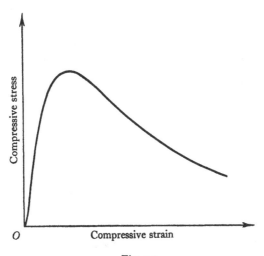

Fig. 4.1

In fig. 4.2 is shown a rectangular section of width b, and depth d to the
tensile reinforcement of total area A_s. If this cross-section is subjected to
a pure bending moment M, then, as the value of M is increased, some
portion of the cross-section will pass out of the elastic range. For an
economical 'under-reinforced' section, that is, one in which the steel
stress is relatively higher (compared with the yield stress of the material)
than the concrete stress, the steel reinforcement will yield first, and it will
be assumed that it does so at constant stress σ_0.

Making the conventional assumption of plane sections remaining plane
on average, then the stresses after yield will be as shown in fig. 4.2(c).
The distribution of stress in the concrete is, in effect, a reproduction of
the stress–strain curve of fig. 4.1. The force balance across the cross-

section shows that the total concrete force C in fig. 4.2 (c) must equal the force in the steel, $\sigma_0 A_s$. As the bending strain is increased, therefore, the shape of the 'stress block' on the concrete will change, but the value of C will remain constant. At any particular state of strain the value of the bending moment corresponding to fig. 4.2 (c) may be calculated; due to the falling characteristic of fig. 4.1, the value of the bending moment will reach a maximum corresponding to the ultimate moment of resistance of the cross-section.

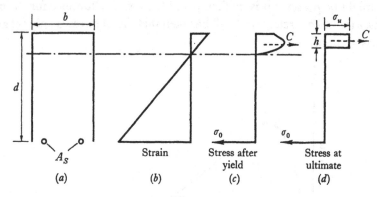

Fig. 4.2

Figure 4.2 (d) shows the conventional rectangular stress block used to calculate the ultimate moment of the section; the value of σ_u is often taken as two-thirds of the nominal crushing strength as observed in a cube test. In fact, as will be seen below, the precise value taken for σ_u affects only slightly the resulting bending moment. Using the symbols in fig. 4.2 (d), the balance of forces gives

$$C = \sigma_0 A_s = \sigma_u bh,$$

or
$$h = \frac{\sigma_0}{\sigma_u} \frac{A_s}{b}, \tag{4.1}$$

and the value of the bending moment is given by

$$M = \sigma_0 A_s(d - \tfrac{1}{2}h). \tag{4.2}$$

As a numerical example, suppose that the reinforcement is 1 per cent (i.e. $A_s = $ 0·01bd), that $\sigma_0 = $ 250N/mm², and that $\sigma_u = $ 20 N/mm². From equation (4.1), the value of h is 0·125d, so that, from (4.2),

$$M = \text{0·9375}\sigma_0 A_s d. \tag{4.3}$$

62

It should be noted that the depth h of uncracked concrete is only $\frac{1}{8}d$, as sketched roughly in fig. 4.2(d). The ultimate moment from equation (4.3) is some 6 per cent less than that corresponding to a full lever arm of depth d.

This point may be emphasized by repeating the calculation for values of σ_u of 30 and 15 N/mm²; the respective lever arms are 0·958d and 0·917d. The value of the ultimate moment of resistance is extraordinarily insensitive to the assumed value of the stresses in the concrete. It will be appreciated that the value of the ultimate moment is also insensitive to the precise form of the assumed distribution of stress across the section, which justifies the use of the rectangular block of fig. 4.2(d). The bending moments corresponding to figs. 4.2(c) and (d) will have almost identical values.

These arguments for an under-reinforced section indicate that, once yield has started in the steel reinforcement, the bending moment stays sensibly constant as the strain is increased, and considerable rotation can take place at such a 'hinge' before any fall in moment occurs. Effectively, the yielding steel controls the behaviour of the cross-section, and confers its own basic ductility on the section as a whole.

4.2 Bending with axial load

On the assumption that some constant limiting stress σ_u can be assigned to the concrete yielding in compression, it is a simple matter to compute the yield surface for the rectangular section of fig. 4.2(a) subjected to both axial load and bending moment. However, the results cannot be used with as much confidence as those for simple bending; it is to be expected that there will be some difference between calculated and observed behaviour. Reinforced concrete is a very variable material, and the value of the 'yield stress' σ_u will vary from member to member, and even within a cross-section. In the case of pure bending it was shown that the 'full plastic moment' corresponding to the stress distribution of fig. 4.3(d) depends only weakly on the actual value of σ_u. This will no longer be true in the presence of large compressive axial loads; in the limit, for example, the 'squash load' of the cross-section will be almost directly proportional to the value of σ_u.

Despite these reservations, a yield surface will be derived for the rectangular cross-section of fig. 4.2(a). In fig. 4.3(a) to (d) are shown the cases that must be considered, in which the zero-stress axis lies at various levels in the cross-section. For small values of the axial load P, the distribution of fig. 4.3(a) is merely a modified form of fig. 4.2(d); as the compressive axial load is increased, so the zero-stress axis moves progressively down the cross-section. The calculations will be referred to the centre-line of the cross-section; that is, the bending moment acts in the presence of an axial load whose point of application is the centre of the rectangle $b \times d$. (For simplicity, no cover is taken for the steel; it is easy to modify the calculations to allow for a finite depth of concrete below the steel level.)

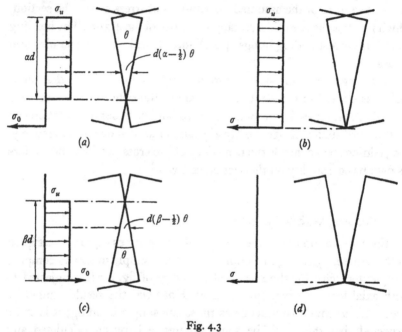

Fig. 4.3

For the stress distribution of fig. 4.3(a), the value of the axial load P (*tensile* positive) is given by

$$P = \sigma_0 A_s - \sigma_u \alpha b d$$
$$= b d \sigma_u \left[k \frac{\sigma_0}{\sigma_u} - \alpha \right], \tag{4.4}$$

and the value of the bending moment M is given by

$$M = \sigma_0 A_s \frac{d}{2} + \sigma_u \alpha \frac{b d^2}{2} (1 - \alpha)$$
$$= \tfrac{1}{2} b d^2 \sigma_u \left[k \frac{\sigma_0}{\sigma_u} + \alpha (1 - \alpha) \right]; \tag{4.5}$$

64

in these expressions, k is the ratio A_s/bd (e.g. $k = 0.01$ for a reinforcement area of 1 per cent).

Equations (4.4) and (4.5) plot as the curve AB in fig. 4.4, as α is allowed to vary from zero to unity. It will be seen that the largest value of M occurs for $\alpha = \frac{1}{2}$, when the top half of the concrete is yielding. For $\alpha = 1$, the zero-stress axis lies at the level of the reinforcement, and the net axial load at point B in fig. 4.4 is the difference between the crushing strength of the concrete and the yield strength of the reinforcement. The compressive axial load can be increased in value, with the zero-stress axis

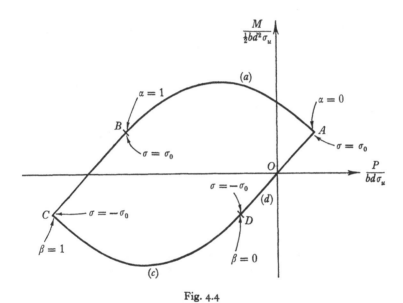

Fig. 4.4

remaining at the reinforcement level, fig. 4.3(b), while the stress in the steel changes progressively from yield in tension to yield in compression. If σ is the value of the stress in the steel,

$$P = bd\sigma_u\left[k\frac{\sigma}{\sigma_u} - 1\right],$$
$$M = \tfrac{1}{2}bd^2\sigma_u\left[k\frac{\sigma}{\sigma_u}\right]. \tag{4.6}$$

Equations (4.6) plot as the straight line BC in fig. 4.4. When, at C, the whole section is yielding in compression, the axis of rotation of the hinge (that is, the zero-stress axis) can move off the bottom of the section as in fig. 4.5 (a). Eventually an axial movement only will occur at the hinge, corresponding to the axis of rotation lying at infinity, fig. 4.5 (b). The axis of rotation will reappear above the beam, first at infinity and then in the general position of fig. 4.5 (c), and each of the three configurations in fig. 4.5 corresponds to the single point C in fig. 4.4. Point C is, of course, a vertex of the yield surface, and the deformations sketched in fig. 4.5 illustrate the indeterminacy of the deformation at such a vertex.

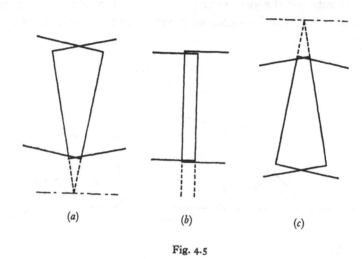

(a)   (b)   (c)

Fig. 4.5

The portion CD of the yield surface in fig. 4.4 corresponds to the stress distribution of fig. 4.3(c). The parametric equations are

$$P = bd\sigma_u\left[-k\frac{\sigma_0}{\sigma_u}-\beta\right],$$

$$M = \tfrac{1}{2}bd^2\sigma_u\left[-k\frac{\sigma_0}{\sigma_u}-\beta(1-\beta)\right].$$

(4.7)

Finally, the zero-stress axis becomes 'locked' again at the level of the reinforcement, fig. 4.3 (d), and the stress in the steel changes progressively from yield in compression to yield in tension along the portion DA of fig. 4.4. The equations for the straight line DA are

$$P = bd\sigma_u\left[k\frac{\sigma_0}{\sigma_u}\right],$$

$$M = \tfrac{1}{2}bd^2\sigma_u\left[k\frac{\sigma_0}{\sigma_u}\right].$$

(4.8)

66

The complete yield surface, fig. 4.4, is skew-symmetric. The normality rule must be obeyed; that is, if axes of deformation (in this case, extension of the centre-line and rotation at the hinge) are superimposed on the force axes (P and M), then the normal to the yield surface gives the relative proportions of the deformations. For example, the gradient of the curve AB in fig. 4.4 is given from equations (4.4) and (4.5) as

$$\frac{dM}{dP} = \frac{\frac{1}{2}bd^2\sigma_u(1-2\alpha)}{(-bd\sigma_u)} = -d(\tfrac{1}{2}-\alpha).$$
(4.9)

Thus the deformations at the hinge must be related by

$$\frac{\theta}{\delta} = \frac{1}{d(\tfrac{1}{2}-\alpha)},$$
(4.10)

which is, of course, a geometrical relationship given directly from fig. 4.3(a).

Along portion AB of the yield surface in fig. 4.4 there is a unique relationship between the loads causing full plasticity and the consequent deformations at the hinge. Along BC and DA, which are flats of the yield surface, the ratio of deformations is fixed (i.e. $\delta = -\tfrac{1}{2}d\theta$), but there is a range of values of M and P to give full plasticity; this corresponds to the zero-stress axis lying at reinforcement level in fig. 4.3(b) and (d). At the vertices A and D, as already noted, there are unique values of P and M, but the zero-stress axis can occupy an infinite number of positions, corresponding to the 'fan' at vertex C of fig. 1.8.

The yield surface of fig. 4.4 is plotted to scale in fig. 4.6 for the cases of 2 per cent, 1 per cent, and zero reinforcement. This last case is of particular importance in the analysis of masonry construction (see section 4.4 later). The 'nesting' of the yield surfaces is a consequence of the safe theorem. That is, any given combination of P and M which causes full plasticity for the section with, say, 1 per cent reinforcement will be, *a fortiori*, an acceptable combination of P and M for the stronger section with 2 per cent reinforcement; no portion of the yield surface for the weaker section can lie outside the yield surface for the stronger section.

4.3 The collapse of simple frames

In the analysis of plane frames by simple plastic theory, using the techniques presented in vol. 1, the effects of axial load (and of shear force) were introduced merely by modifying the value of full plastic moment at the hinge points. That is, it was not found necessary to use a flow rule (as it was for the case of combined bending and twisting, chapter 2) in order to determine collapse loads. This is because the *type* of collapse mechanism for a plane frame remains the same whether or not the effects of axial load are taken into account. For a reinforced-concrete frame, for example, the axes of the plastic hinges will not lie on the centre-line of the frame, but will be displaced to positions such as those sketched in

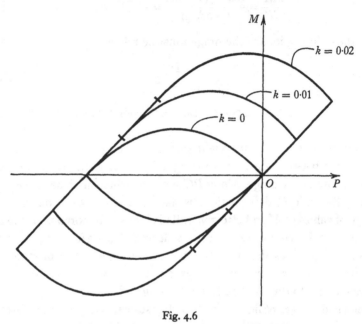

Fig. 4.6

fig. 4.3. Such displacements will alter slightly the geometry of the collapse mechanism; whether or not these small geometrical changes can be ignored, a frame having a number R redundancies will still require $(R+1)$ hinges for a regular collapse mechanism.

Now the bending moment M and axial force P at each hinge position can be determined in general in terms of the R unknown quantities. Further, at each hinge a relation between M and P may be established, either analytically, or by reference to some known plot for the cross-section such as fig. 4.4. Thus, as usual, the $(R+1)$ such relations at the hinges may be used to eliminate the R unknown quantities, leaving a single collapse equation. Again, as usual, there is the possibility of the formation of an incomplete mechanism, for which part of the frame remains redundant.

These arguments are, in fact, identical with those for the simple case in which axial load is ignored, and confirm that allowance for axial load, while necessarily making the working more difficult, does not alter the nature of the calculations. The equations cannot usually be solved explicitly, and a trial and error numerical method may have to be used.

As a very simple example, the arch of fig. 4.7 was shown in vol. 1 to collapse at a load

$$W = 16\frac{M_p}{l}, \qquad (4.11)$$

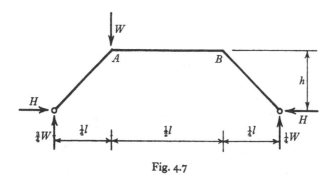

Fig. 4.7

where M_p is the value of the uniform full plastic moment of the cross-section. Equation (4.11) will be modified to allow for the effect of axial load. The doubly-reinforced section of fig. 4.8 has the yield surface of fig. 4.9; cover to the reinforcement has again been ignored. The relevant portion, (a), of the curve has equations

$$P = -bd\sigma_u(\alpha),$$
$$M = kbd^2\sigma_0\left\{1 + \frac{\sigma_u}{2k\sigma_0}\alpha(1-\alpha)\right\}. \qquad (4.12)$$

Numerically, the factor $\sigma_u/2k\sigma_0$ will be taken as 5, and the values of P and M will be taken as

$$P = -2000\alpha \text{ kN},$$
$$M = \pm 100\{1 + 5\alpha(1-\alpha)\} \text{ kN m}, \qquad (4.13)$$

where the appropriate sign must be taken for M according as the hinge is sagging or hogging.

The collapse mechanism of the arch of fig. 4.7 will involve hinges at A and B, fig. 4.10(a). The moments M_A and M_B at these hinges will not

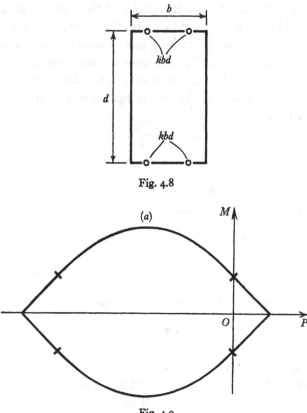

Fig. 4.8

Fig. 4.9

in general be equal, since the thrusts in the sloping legs of the arch are unequal. On the usual assumption that the depth d of the arch is small compared with the overall dimensions, the work equation for the collapse mechanism gives

$$W\left(\frac{l}{4}\theta\right) = M_A(-2\theta) + M_B(2\theta),$$

or

$$W = \frac{8}{l}(M_B - M_A). \tag{4.14}$$

The horizontal component H of the abutment thrust may be found by simple statics, or, equivalently, by considering some suitable virtual mechanism, such as that of fig. 4.10(b), which gives directly

$$W\left(-\frac{l}{4}\theta\right) + H(2h\theta) = M_A(\theta) + M_B(\theta),$$

or

$$H = \frac{1}{2h}(3M_B - M_A), \tag{4.15}$$

70

using the value of W given by equation (4.14). This value of H may be combined with the vertical reactions at the abutments to give the thrusts P_A and P_B in the two legs of the arch.

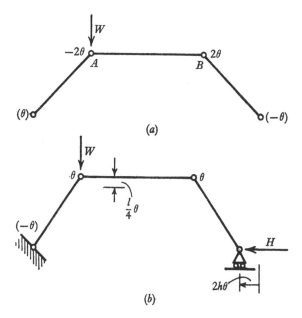

Fig. 4.10

For an arch having $l = 8$m, $h = 2$m, equations (4.14) and (4.15) become

$$W = M_B - M_A,$$
$$H = \tfrac{1}{4}(3M_B - M_A),$$

and the thrusts P_A and P_B become

$$P_A = -\frac{1}{\sqrt{2}}(H + \tfrac{3}{4}W),$$

$$P_B = -\frac{1}{\sqrt{2}}(H + \tfrac{1}{4}W).$$

(4.16)

These four equations must be solved together with equations (4.13) for the yield surface.

Starting from the solution ignoring the effect of axial load, i.e. assuming $P_A = P_B = 0$, equations (4.13) and (4.16) may be examined in the order given, and lead to

$$\alpha_A = 0, \quad \alpha_B = 0, \quad M_A = -100, \quad M_B = 100, \quad W = 200,$$
$$H = 100, \quad P_A = -177, \quad P_B = -106.$$

(4.17)

The new values of P_A and P_B lead to

$$\alpha_A = 0.088, \quad \alpha_B = 0.053, \quad M_A = -140, \quad M_B = 125, \\ W = 265, \quad H = 129, \quad P_A = -232, \quad P_B = -138. \quad \left.\right\} \quad (4.18)$$

Three more cycles lead to the steady values

$$M_A = -155, \quad M_B = 135, \quad W = 290. \quad (4.19)$$

Thus the effect of allowing for axial load in this particular problem is to increase the collapse load of the arch from 200 to 290 kN.

From the way in which the calculations have been set out, it will be appreciated that a solution is obtained at each stage which satisfies the yield condition (equations (4.13)) and equilibrium (the first two of equations (4.16)). Thus the resulting value of W must be a lower bound to the correct collapse value, and this is true at each stage of the calculations.

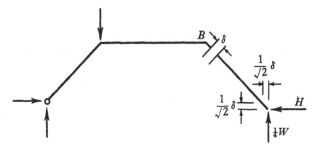

Fig. 4.11

More complex arches and frames may be solved by the same sort of numerical computation. A trial collapse mechanism will furnish a set of axial thrusts in the structure, whose values may be used to modify the full plastic moments at the hinge positions. These new values of full plastic moment may be used in turn to derive a new set of axial thrusts, and a very few iterative cycles will in general lead to the correct answers for the assumed mechanism. The whole frame must then be checked to confirm that a hinge is not formed elsewhere, and a modified mechanism must be examined if necessary. In all these calculations, the equation of virtual work will give the simplest derivation of the necessary equilibrium equations; figs. 4.10 were used to obtain equations (4.14) and (4.15). Similarly, the thrusts at internal hinges may be found by examining virtual displacements involving *extensional* (rather than rotational) deformations at the hinges. The last of equations (4.16), for example, follows directly from the virtual deformation of fig. 4.11.

4.4 The masonry structure

It is of interest to discuss here the behaviour of material which is incapable of carrying tensile stress, and which is also unreinforced. Dry sand is an obvious example of such a material; soil mechanics is outside the scope of this book, but the whole subject has benefited greatly by the use of ideas stemming from simple plastic theory†. Masonry, assembled dry or with weak mortar, is another example, and the masonry structure has also been examined‡ in the light of the plastic theorems.

The yield surface for masonry has already been sketched; it is the innermost curve ($k = 0$) of fig. 4.6, and consists of two parabolic arcs. The yield surface lies wholly on the negative side of the P axis; in accordance with the assumption that tensile stresses cannot be carried, some compressive load P must be present if the material is to act structurally. In fact, the mean compressive stresses in traditional masonry construction, such as voussoir arch bridges, or cathedrals, have extremely low values; it is very rarely that they attain as much as 10 per cent of the crushing strength of the material. In these circumstances, the approximation may be made that, relative to the working values of compressive stress the ultimate strength of the material is infinite. The parabolic yield surface, shown with broken lines in fig. 4.12, will then be replaced by the two straight lines; a low general stress level implies that attention is confined to a region such as that shown hatched.

The significance of the two straight lines may be seen by reference to fig. 4.13. Here a hinge is just forming between two adjacent blocks of masonry (e.g. voussoirs), the thrust P being transmitted at the point of contact (and producing, locally, infinite stresses). Referred to the centre line of the voussoirs, the bending moment M may be written

$$M = \pm P(\tfrac{1}{2}d),\qquad(4.20)$$

where the sign of M again depends on whether the hinge is hogging or sagging. The two lines (4.20) are the two straight lines in fig. 4.12. For comparison, setting $k = 0$ in equations (4.4) and (4.5) leads to the expressions

$$\left.\begin{aligned}P &= -\alpha P_0,\\ M &= -P(\tfrac{1}{2}d)(1-\alpha),\end{aligned}\right\}\qquad(4.21)$$

† A. N. Schofield and C. P. Wroth, *Critical state soil mechanics*, London (McGraw-Hill), 1968.

‡ J. Heyman, The stone skeleton, *Int. J. Solids Structures*, **2**, 1966; The safety of masonry arches, *Int. J. mech. Sci.* **11**, 1969.

which are the parametric equations for the upper parabolic arc in fig. 4.12.

Once again, the general properties of the yield surface apply to the particular surface of fig. 4.12. Confining attention to the hatched area, a point lying within this area represents a combination of P and M that can be carried safely by the masonry. That is, the point corresponds to the line of thrust lying within the total depth d, fig. 4.13, so that no hinging

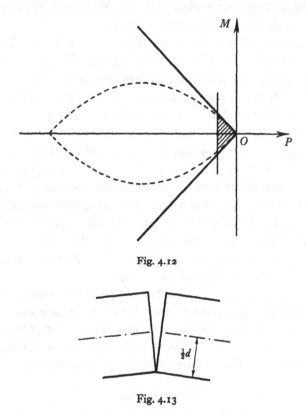

Fig. 4.12

Fig. 4.13

action occurs. As mentioned above, a point lying on the boundary corresponds to the formation of a hinge. Points outside the boundary represent an impossible state; the line of thrust cannot lie outside the masonry.

The assumption that stone has no tensile strength is a 'safe' assumption. If the mortar used in an actual structure is capable of transmitting some tension, or if the stones are interlocked so that, again, tensile forces can be carried, then the yield surface for the real structure *must* lie outside that for the structure in which tensile forces are ignored. Any satisfactory solution for the simpler structure will then certainly be satisfactory for the real structure. However, this safe assumption is counteracted by the second assumption that stone has infinite compressive strength, by which the parabolic yield surface is replaced by two straight lines. The errors here will usually be small, but the calculations can in any case again be made safe by means of a simple artifice.

For small values of mean compressive stress, the straight lines of fig. 4.12 can be replaced by straight lines of smaller slope. For example, putting $\alpha = 0 \cdot 1$ into equations (4.21) suggests that equation (4.20) could be replaced by

$$M = \pm 0 \cdot 9 P(\tfrac{1}{2}d), \qquad (4.22)$$

if it were thought that P/P_0 was likely to be less than $0 \cdot 1$ throughout the structure. All calculations could then be referred to the inner pair of lines in fig. 4.14. Since this artificial yield surface lies within the true parabolic surface, the calculations have been made 'safe'. In this example, the use of equation (4.22) may be thought of as referring to an artificial structure whose overall depth at every section has been reduced to 90 per cent of the real depth d; the assumption that the actual mean stresses are less than 10 per cent of the crushing strength can of course be checked *a posteriori*.

In the following discussion no further mention will be made of this device; it will be assumed either that the stresses are so very low that it is not needed, or that the structure has already been 'shaved down' to a suitable fraction of its original depth before starting the analysis. Further, it will be assumed that no other mechanism of failure, other than hinging, can occur; that is, for example, it will be taken that compressive forces between adjacent portions of masonry are sufficient for consequent frictional forces to prevent any form of sliding failure. Under these conditions, the three master statements of plastic analysis, those of mechanism, equilibrium, and yield (statements (3.50) in vol. 1), may be 'translated' to apply to the masonry structure.

The *mechanism condition* remains unchanged; sufficient hinges must form to turn the structure into a mechanism at collapse. Thus, for a voussoir arch, a possible arrangement of hinges, alternately opening and closing (i.e. formed in the intrados and the extrados), is shown in fig. 4.15. The corresponding four-bar chain is shown by broken lines connecting the hinge points. Note that the arch is a structural form having (in conventional terms) three basic redundancies, so that the formation of four hinges at collapse is to be expected.

There are many structures, of which two are sketched in fig. 4.16, for which no pattern of hinges will lead to a possible mechanism. Thus in the low-rise arch of fig. 4.16(a), no arrangement of hinges in the extrados and intrados of the arch, similar to that of fig. 4.15, can be devised to give

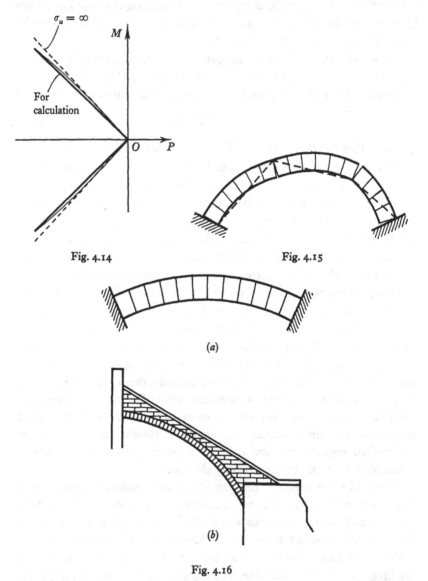

Fig. 4.14

Fig. 4.15

(a)

(b)

Fig. 4.16

a four-bar chain; the same is true of the idealized flying buttress of fig. 4.16(b). (This flying buttress is of a standard thirteenth-century shape.) Since the mechanism condition cannot be satisfied, no further analysis is necessary; providing the abutments are firm, the actual masonry structures of fig. 4.16 cannot collapse *under any loading system*. In practical terms this means that the applied loads must be increased enormously until the yield surface is severely modified by crushing of the material. At such large loads the simplifying assumptions leading to the two straight lines of fig. 4.12 break down, and the actual yield surface must be used to obtain an exact analysis of the collapse behaviour.

Continuing the discussion for light loads, the second of the three master conditions, that of *equilibrium*, will be satisfied by the construction of a line of thrust for the structure which is in equilibrium with the applied loads, including self-weight. The *yield condition* will be satisfied if such a line of thrust lies everywhere within the masonry. There is no question here of a 'middle-third' rule, which is essentially an elastic concept. The fact that the thrust line merely lies within the masonry ensures that forces can be transmitted by purely compressive action from one section to the next.

It will be appreciated that, in general, the thrust line for a particular structure under given loading cannot be determined uniquely from equilibrium considerations. This corresponds to the fact that a redundant structure possesses an infinite number of equilibrium states, which can be derived by varying the values of the redundant quantities. The *safe theorem* states that if any one line of thrust can be found for given loading which lies wholly within the masonry, then this demonstrates that the structure will not collapse under those loads. Similarly, the usual corollaries of the safe theorem may be applied to masonry. For example, any small imperfections, such as settlements or spread of abutments which do not sensibly affect the overall geometry of the structure, cannot of themselves promote collapse. Such imperfections may well cause cracks to appear in the masonry; that is, certain hinges may form, but these will be insufficient in number to transform the structure into a mechanism. A thrust line satisfying the safe theorem for a structure before settlement will also satisfy the theorem after settlement.

This duality of a mechanism and an equilibrium approach may be amplified. It was demonstrated, negatively, that a structure such as that of fig. 4.16(a) could not collapse, since no possible arrangement of hinges to form a mechanism could be found. The stability of the structure can be demonstrated alternatively, and positively, by using the safe theorem. Thus, in fig. 4.17, an arbitrarily positioned point load is shown equilibrated by a thrust line lying wholly within the masonry. Straight lines can be drawn in this way from every point on the upper surface of the arch to the abutments; this is a complete demonstration that, providing the abutments hold, there is no combination of downward vertical loads for which it is not possible to find a line of thrust lying wholly within the masonry.

As a final problem in the plastic analysis of masonry, the parabolic arch will be investigated. The arch will be supposed to be subjected to a heavy uniformly distributed load, and to a light travelling point load, and the arch will be designed to have the minimum possible thickness to just

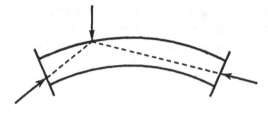

Fig. 4.17

contain a thrust line within the masonry. (The actual arch could then be built with a thickness say λ times this theoretical minimum, in which case it would incorporate a 'geometrical' factor of safety of value λ. A suitable value of λ might lie between 5 and 10.)

The reference to a 'heavy' dead load above was made to simplify the working; it will be assumed that the abutment thrust is not altered by the 'light' live loading, so that the limiting hinge moments ($M = \pm P(\frac{1}{2}d)$ from equation (4.20)) are known before the calculations start. To simplify the working further, it will be assumed that the value of the hinge moment is the same at all cross-sections. In a real arch the hinge moment will be related to the depth of the voussoirs at any section and to the local value of the thrust; an iterative process may be used, as in section 4.3 above, to determine the collapse conditions.

Figure 4.18(a) shows the parabolic arch, span $2l$ and rise h, subjected to a dead load W, uniformly distributed, and to a live load P distant xl from the centre-line of the arch. The line of thrust for the dead load is itself parabolic, and will be taken as coinciding with the centre-line of the arch. By the usual simple theory of arches, therefore, the horizontal component H of the abutment thrust is given by

$$H = \frac{Wl}{4h}. \qquad (4.23)$$

A possible collapse mechanism for the arch is shown in fig. 4.18(b), in which hinges are formed under the live load, at the two abutments, and at some point, distant yl from the centre-line of the arch, between the live load and the right-hand abutment. When the live load is close to centre span, an alternative mechanism can occur in which the hinge moves away

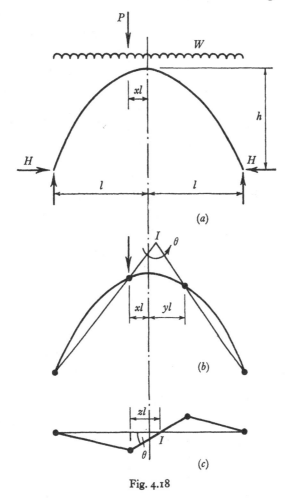

Fig. 4.18

from the left-hand abutment; this is discussed below (fig. 4.19). The mechanism of fig. 4.18(b) is the most critical for the parabolic arch.

With the assumption of a constant hinge moment, the problem posed by fig. 4.18(b) becomes an exercise in simple plastic theory, albeit complicated by some slightly difficult geometry. A small motion of the collapse mechanism, expressed in terms of a rotation θ about the instantaneous centre I marked in fig. 4.18(b), will lead to a work balance for the collapsing arch, from which a value M_p of 'full plastic moment' may be calculated. As will be seen below, this value of M_p may be used to calculate directly the required thickness of the arch, that is the thickness that will just contain the thrust line.

Now the hinges of fig. 4.18(b) lie on the centre-line of the parabolic arch, this parabola also being the line of thrust for the uniformly distributed load W. Any small inextensional deformation of such a thrust line must lead, by the equation of virtual work, to zero contribution from the corresponding loads; that is, the work balance for the hinge mechanism of fig. 4.18(b) will contain loading terms involving only the live load P. This is a general property for the analysis of any real arch of arbitrary shape, and is not confined to the artificial parabolic arch under discussion; considerable simplification results by referring calculations to a dead-load thrust line, as will be illustrated later.

The hinge discontinuities of fig. 4.18(b) may be determined from the 'projected' mechanism of fig. 4.18(c); the location of the instantaneous centre I is an intermediate step in the calculations, but the parameter z can clearly be determined in terms of x and y (the position yl of the hogging hinge has not yet been found). The work equation then gives

$$M_p = \frac{Pl}{2}(1-x)\frac{\{x+y(1-x)-y^2\}}{\{(2-x)+y(1-x)-y^2\}}. \qquad (4.24)$$

In this equation, the position of the load is assumed fixed; that is, the value of x is known. The unsafe theorem states that the value of M_p must be maximized with respect to y; this condition gives

$$y = \tfrac{1}{2}(1-x). \qquad (4.25)$$

That is, the hinge point lies exactly half-way (horizontally) between the loading point and the right-hand abutment. Substituting this value of y into equation (4.24),

$$M_p = \frac{Pl}{2}(1-x)\frac{(1+x)^2}{(3-x)^2}. \qquad (4.26)$$

Equation (4.26) can now in turn be investigated to find the worst position of the live load. The value of M_p is a maximum for

$$x = 5-\sqrt{20} = 0.528,$$

that is, for the load at about quarter span, and the maximum value is

$$[M_p]_{\text{max}} = (\tfrac{1}{2}Pl)(5\sqrt{5}-11) = 0.0902Pl. \qquad (4.27)$$

Before discussing this analytical problem more fully, the results may perhaps be illuminated by inserting numbers corresponding to a hypothetical parabolic bridge arch. Making the calculations for a unit (metre) width of bridge of span 24 m and rise 3 m (i.e. $l = 12$, $h = 3$), the dead load W will be taken as 1500 kN uniformly distributed, and the live load P as 100 kN. From equation (4.23), the basic value of the abutment thrust is $H = Wl/4h = 1500$ kN, and no correction to the value of H will be made in this simple example to allow for the effect of the live load. The value of the required 'full plastic moment' is given from equation (4.27) as $M_p = (0.0902)(100)(12) = 108.2$ kN m. Now the value of the hinge moment is given also by the product (thrust × eccentricity), so that the quantity $\frac{1}{2}d$ in fig. 4.13 is given by (within the approximations of the present calculations)

$$1500(\tfrac{1}{2}d) = 108.2, \tag{4.28}$$

that is $d = 0.144$ m. Thus an actual bridge of this type whose depth d was 1 m would have a geometrical factor of safety of about 7. The mean stress level in the actual bridge would be about $(1.5)/(1 \times 1) = 1.5$ MN/m², on simply dividing the thrust H by the cross-sectional area of the arch ring. The crushing strength of a medium sandstone is about 40 MN/m², which confirms the generally low stress levels in such structures, and justifies the use of the simplified straight-line yield surface, fig. 4.12.

The 'work method' which led to equation (4.24) is based on the established concepts of simple plastic theory, and the subsequent steps to determine the maximum value of M_p were straightforward. For the general problem of a masonry arch, however, particularly one of more complex geometry, the method is not really convenient, and the solutions may be found most easily by drawing. Since the dead load W in fig. 4.18(a) for the parabolic arch did not appear in the expressions for M_p, equations (4.24)ff., it is sufficient to determine the effect of the live load acting alone.

In order to find the value of M_p at the four hinge points of fig. 4.18(b), therefore, the thrust line of fig. 4.19(a) may be examined. Since the bending moment in the arch is given by the product of the horizontal component of the abutment thrust and the vertical eccentricity of the thrust line, fig. 4.19(a) shows the thrust line positioned to give four equal eccentricities e. It is clear that the value of e can be found quickly by drawing for any position of the load. (The fact that the hogging hinge lies half way between the loading point and the right-hand abutment is immediately obvious from fig. 4.19(a) as one of the basic properties of a parabola).

It is important to be clear that fig. 4.19(a) represents a subsidiary and not the main problem. That is, having determined the value of e, the horizontal component H' has nothing to do with the corresponding value H for the full dead loading. The product $H'e$ leads (numerically) to equation (4.26) for the value of M_p; the investigation of different positions of the live load P will reveal the worst case, for which the maximum value of M_p is given by equation (4.27). Thus, having determined the result of

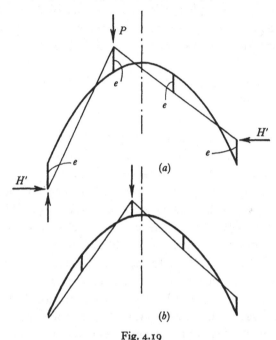

Fig. 4.19

equation (4.27) by graphical methods rather than analytically, the work may proceed as before to find the minimum depth of arch ring necessary just to contain the dead plus live-load thrust line.

It may be noted that as the live load P in fig. 4.19(a) approaches mid-span, the hinge at the left-hand abutment moves away from that abutment, and the collapse mechanism forms as in fig. 4.19(b). A schematic plot of the value of M_p required for a unit load placed anywhere on the arch is shown in fig. 4.20. A central point load is, as might be expected, not so critical as a load placed at about quarter span.

An actual arch will not have a parabolic centre-line, nor will the dead-load line of thrust coincide with the arch centre-line. However, the techniques illustrated by the above calculations can be extended to deal with an arch of any given centre-line and of variable depth (as fig. 4.21 (*a*)) subjected to any combination of known dead and live loading. The first step in such an analysis is to determine a dead-load line of thrust lying as close as possible to the centre-line of the arch, as shown schematically in fig. 4.21 (*b*). The real arch is then imagined to be shrunk towards its

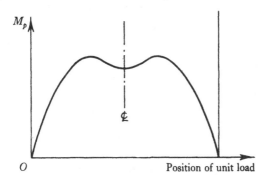

Fig. 4.20

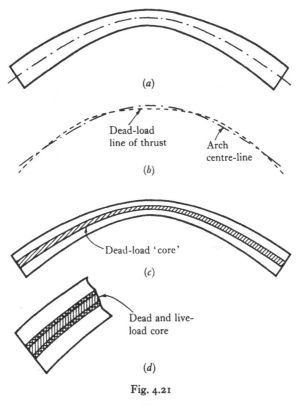

(*a*)

Dead-load
line of thrust
                        Arch
                        centre-line

(*b*)

Dead-load 'core'

(*c*)

Dead and live-
load core

(*d*)

Fig. 4.21

83

centre-line, the shape and relative depths being preserved, and a solution is sought to the problem of finding the minimum possible thickness of such an arch. The shaded 'core' in fig. 4.21(c) would just contain the dead-load line of thrust of fig. 4.21(b), and represents the solution to the first stage of the problem.

The next step in the solution refers the calculations not to the actual arch centre-line, but to the dead-load line of thrust determined in stage one. A fictitious arch whose centre-line coincided with this dead-load line of thrust could, theoretically, have zero thickness (admitting infinite stresses) and still support the given dead loads, but some finite thickness is required if the arch is to be stable under live loads. This required thickness for the fictitious arch can be found by the graphical technique illustrated in fig. 4.19 for the parabolic arch; by referring the calculations to the dead-load centre-line, only the live loads enter into the problem at this stage. (Naturally, a complex pattern of given live loading can be analysed in this way, not only the simple case of a single point load illustrated in fig. 4.19.) For an arch of variable depth, the quantities e in fig. 4.19(a) would not be made all equal, but would have values in proportion to the depth of the real arch at the corresponding cross-sections.

At the end of this second stage of the calculations, therefore, the minimum thickness of fictitious arch would have been established which has as centre-line the dead-load line of thrust and which can carry the given live loads. This fictitious arch can be 'fitted' back into the real arch profile as shown in fig. 4.21(d); the dead-load 'core' has been thickened to contain the fictitious arch. Thus, finally, a geometrical factor of safety for a given arch can be established by dividing the actual depth of the arch by the depth of the hatched core in fig. 4.21(d).

4.5. Reinforced-concrete arches

The two simplifications which were made in the above discussion of the behaviour of masonry were that stone has no tensile strength, and that the compressive stresses were so low that the ultimate compressive strength of the stone could be regarded as infinite. Voussoir construction was implied, but this restriction is unnecessary; any masonry structure whose yield surface can be approximated as in fig. 4.12 can be discussed by the methods of the previous section. Thus a complete structure built of small pieces of stone (as, for example, a Gothic cathedral) will behave according to the basic laws of plastic theory, providing that the specific assumptions (low compressive stresses, no tensile strength) are obeyed, and providing that the general assumption of all simple structural theory, that of small deflexions, is also obeyed.

For simplicity in working the examples of masonry arches, it was further assumed that the horizontal component of the abutment thrust was not affected by the magnitude and position of the live loads. Again, this is an unnecessary restriction made only in the interests of clarity of exposition; the basic problem is the construction of the dead plus live-load core of fig. 4.21 (d), and this can be done graphically using say half a dozen positions of live loading.

The fundamental problems of plastic design in reinforced concrete were illustrated by the simple arch of fig. 4.7, using the yield criterion of fig. 4.9. With a closed yield surface of this type there is no question of a structure of infinite strength; an arch of exactly the right shape to carry the dead loading without bending, and whose depth has been adjusted to give a constant stress throughout, will fail simultaneously at all sections in pure compression if the dead loads are increased uniformly by a suitable load factor. However, some of the techniques suggested by the analysis of masonry can be employed usefully in the analysis of reinforced concrete.

In particular, the use of a dead-load line of thrust instead of the real centre-line of an arch leads to a much easier analysis of the effect of live loading. It is difficult to obtain analytical solutions to practical problems, but graphical and numerical methods can lead to iterative solutions which converge rapidly. Exactly as for the problem of fig. 4.7, a first trial solution will enable the construction of a thrust line, and axial forces P may be calculated for each critical section. The use of a yield surface such as that of fig. 4.9 (not necessarily the same at each critical cross-section) will lead to values of bending moment at the assumed hinge positions, and the thrust line can be modified to accord with these. This modification will lead to revised values for the thrusts, although the revisions may well be very small for reasonably large values of dead compared with live loading. (Note that the line of thrust can lie outside the cross-section, due to the presence of steel reinforcement.)

These techniques have been explored thoroughly by Franciosi,† who gives sample calculations for a reinforced-concrete arch of span 100 m. It is possible also to take into account the effect of variable repeated

† V. Franciosi, *Scienza delle costruzioni*, vol. 4, *Calcolo a rottura*, Napoli (Libreria Liguori), 1964.

loading, which might lead to the possibility of incremental collapse rather than the safe state of 'shakedown'. These particular topics are outlined with respect to steel frames in chapter 6.

EXAMPLES

4.1 Establish equations (4.12) for the yield surface (fig. 4.9) of the doubly-reinforced rectangular cross-section of fig. 4.8. Plot the yield surfaces (for $\sigma_0/\sigma_u = 10$) for the three cases $k = 0$, 0·01, 0·02.

4.2 The symmetrical cross-section shown is subjected to an axial load P (acting on the centre-line of the cross-section) and to a bending moment M. Plot the yield surface for $\sigma_0/\sigma_u = 10$, $c/d = 0·1$, and $k = 0·01$.

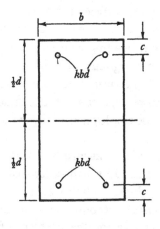

4.3 Repeat example 4.2 for the cross-section shown, reinforced on one side only. Use the same numerical values as in example 4.2.

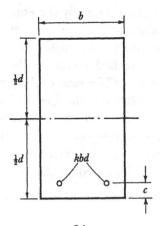

4.4 The reinforced-concrete arch of fig. 4.7 ($l = 8$m, $h = 2$m) is built with *fixed* abutments, capable of developing the full moment of resistance of the cross-section. The relevant portion of the yield surface for all cross-sections is given by equations (4.13). Determine the value of the load W at collapse.

(*Ans.* 510 kN.)

4.5 The derivation of equation (4.25) from (4.24) is not completely straightforward. Examine this derivation, and deduce equations (4.26) and (4.27).

4.6 The arch whose centre-line is shown is made of masonry voussoirs each of constant *vertical* depth d (so that the 'full plastic moment' of the arch is constant). The arch is subjected to two dead loads W, and to a travelling live load P of much smaller magnitude. By considering a plastic collapse mechanism of the type shown in fig. 4.18(b), show that the value of full plastic moment required just to support the loads is given by

$$M_p = \frac{Pl}{2}\left(\frac{1-x^2}{4-x}\right).$$

Deduce that the maximum value of M_p occurs for $x = 4 - \sqrt{15} = 0.127$, and that, correspondingly, $M_p = 0.127Pl$. Deduce that the maximum eccentricity e of the thrust line (cf. fig. 4.19(a)) is given by $e/h = 0.127P/W$. Calculate the geometrical factor of safety if $d = 0.8$ m, $h = 3$ m, $W = 10$ MN, $P = 1$ MN.

(*Ans.* 10.5.)

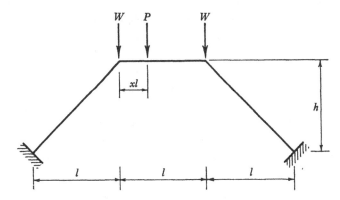

4.7 The centre-line of a symmetrical arch is a circular arc of radius R subtending an angle 2α at the centre. The arch is subjected to dead load which is distributed uniformly horizontally. If the uniform *vertical* thickness of the arch is βR, show that

$$\beta = \frac{1}{4}\frac{(1-\cos\alpha)^2}{(1+\cos\alpha)},$$

if the thrust line due to the dead load is just to be contained within the arch cross-section. Evaluate β for $\alpha = 30°$ and $45°$. (*Ans.* 0.0024, 0.0126.)

5

ELASTIC–PLASTIC ANALYSIS

Simple plastic theory is concerned with the collapse state of structures, and the calculations furnish corresponding values of collapse loads (or load factors), or, alternatively, enable a design to be made to sustain given loads. By its nature the theory gives no information on the deformation of a structure as a whole, or of individual members, beyond the derivation of an ultimate mechanism of collapse. Further, a frame will in general not collapse by a *regular* mechanism for which the whole structure becomes statically determinate; rather, the typical behaviour is to collapse by a *partial* mechanism, which leaves a portion at least of the frame still redundant. For the purpose of assessing collapse loads within the assumptions of simple plastic theory, this behaviour is satisfactory; that is, providing it is *known* (for example) that deflexions are small (in the structural sense of not interfering with the equilibrium equations), and that local and overall stability is assured, then reliance can be placed on the calculated values. The simple theory is, effectively, a rigid–plastic theory, and it is assumed in the first instance that a real structure conforms to that theory.

The designer will wish in practice to make sure that the basic assumptions are satisfied; for example, he will wish to estimate deflexions to ensure that he can indeed apply the simple theory, and to check that overall deformations do not impair any structural function. Similarly, a partial mechanism of collapse may lead to a certain lack of information for the designer; he may not be able, for example, to check the stability of some of the columns. A simple portal frame might collapse in the beam mode of fig. 5.8(b); that this collapse mode is correct may be demonstrated by the use of the safe theorem, that is, by the construction of *any one* distribution of bending moments in the columns such that both equilibrium and the yield condition are satisfied. But the designer is not necessarily interested in a *possible* distribution of bending moments; he will want the *actual* distribution in order that he may check the stability of individual column lengths.

In order to obtain this kind of information, it would seem that elastic theory must be introduced into the analysis, in addition to the simple plastic theory used for the calculation of collapse loads; in fact, an elastic–plastic analysis is required. This means that, instead of the simple rigid–plastic yield condition which states that the bending moments everywhere in a frame must be less than, or at most equal to, the local values of the full plastic moment, the full moment–curvature relationship must be used; in the elastic range, moment is proportional to curvature, and in the plastic range, $|M| = M_p$, fig. 5.1. The relationship of fig. 5.1. is a reasonable approximation for the purposes of calculating deflexions. The neglect of the spread of the plastic zones, that is, the assumption of unit shape factor leading to the two straight lines of fig. 5.1., will lead to a calculated behaviour of a frame that is rather stiffer than it should be.

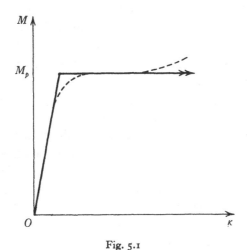

Fig. 5.1

On the other hand, structural mild steel will strain harden at high curvatures, and the full plastic moment will increase above the nominal value M_p. The behaviour corresponding to the two portions of fig. 5.1 shown by broken lines is thus largely self-cancelling.

Much more serious as a source of error is the reintroduction into a problem of plastic analysis of elastic compatibility conditions. In simple plastic theory the deformation equations of the theory of structures are reduced to the vestigial form of the requirement of the formation of a mechanism of collapse. For the simple type of framed structure considered here, the formation of a collapse mechanism makes statically determinate the portion of the frame concerned with the collapse, and the bending moments there are determined uniquely and without reference to compatibility conditions.

By contrast an elastic frame, or those portions of a collapsing frame which do not form part of the collapse mechanism, remain statically indeterminate. There is an infinite range of possible equilibrium states for such structures; that state will be correct which leads to deformations satisfying the compatibility conditions. That is, for example, a fixed-ended beam must deform in such a way that the slopes and deflexions at its ends are zero. Now although curvatures can be calculated with reasonable accuracy from the idealized curve of fig. 5.1, and allowance can be made for a discontinuity in slope should a plastic hinge have formed, it is the elastic boundary conditions on deformations that are so suspect in any attempted analysis of the behaviour of a real structure. A very small rotation at a supposedly clamped end, for example, can lead to a marked relaxation in the value of the bending moment at that end. Similarly, settlement of a foundation, slip in a connexion, or a small dimensional error in fabrication, can lead to extraordinarily large changes in the elastic state of a structure.

It was precisely this kind of consideration which led to the development of the plastic theory of framed structures. It is in a sense not meaningful for a designer to ask for the *actual* state of a redundant elastic structure. As was noted in the closing pages of vol. 1, even if due allowance could be made for all the practical anomalies which might affect the elastic state, the subsequent calculations would still hold only for that particular group of practical imperfections. Any later settlement, slip in a connexion, and so on would in effect generate a new structure about which little elastic information was available.

Despite these remarks, the designer is not relieved of the responsibility of doing the best he can to estimate deflexions and to check the stability of columns. A first and obvious step is to make some sort of elastic calculation, but to view the numerical results of any such computation with some suspicion. It might be that the designer should use, not a *reasonable* distribution of bending moments (which is really what he is accustomed to doing in a conventional analysis), but a *worst possible* distribution given by some other means. In checking column stability, for example, it may be better in a given case to use limits which arise naturally in the collapse analysis rather than to rely on the results of an elastic–plastic calculation. An example of such limits is given in section 5.6 later for the simple rectangular portal frame.

This chapter, then, will be devoted to attempts to estimate the states of those portions of a structure which remain statically indeterminate at collapse, and to assess the deflexions of partially plastic frames. In discussing elastic–plastic states, the equation of virtual work is, as usual, an essential tool.

5.1 Virtual work for elastic–plastic frames

In vol. 1 the equation of virtual work was used to obtain equilibrium statements about frames; that is, the equation was used to generate relationships between values of the bending moments at the various critical sections, whether those values corresponded to full plasticity or not. For this purpose, the actual bending moments in the frame were used as the equilibrium set in the virtual work equation. The compatible sets of displacements were confined to those arising from mechanisms; that is, deformations resulted from straight bar mechanisms, the hinge rotations being the only internal distortions of the frame. The mechanisms were either the actual mechanisms of collapse, in which case the collapse equation could be written by thinking of a 'work balance' and equating the work done by the external loads to the work dissipated in the hinges, or they were virtual mechanisms, not representing the collapse state of a particular frame, but nevertheless furnishing relations between the values of bending moment.

For the purposes of the present chapter, the equation of virtual work will be used the other way round. Instead of combining a real equilibrium state with a set of virtual deformations, the real deformations of the frame will be combined with equilibrium states which have nothing to do with the actual loading on the frame. Since elastic–plastic deformations are under investigation, allowance must be made in the virtual work equation for elastic distortions of the members as well as for hinge discontinuities. At any particular stage of loading (which may be the collapse state) the actual bending moment distribution M throughout the frame will imply elastic curvatures $\kappa = M/EI$ at sections where the full plastic moment has not been reached, and hinge rotations ϕ at sections where $|M| = M_p$. The deformation statement to be used in the equation of virtual work is then:

Plastic hinge discontinuities ϕ and elastic curvatures M/EI
lead to deflexions δ of the frame. $\hspace{2cm}$ (5.1)

As usual, the deflexions δ are measured at the loading points of the frame.
The equilibrium statement to be used is:

Bending moments M^* are in equilibrium with external loads W^*. $\hspace{0.5cm}$ (5.2)

Statements (5.1) and (5.2) can be combined by the equation of virtual work to give

$$\Sigma W^* \delta = \Sigma M^* \phi + \int M^* \frac{M}{EI} ds. \tag{5.3}$$

On the left-hand side, the summation extends over all the concentrated loads W^*, and $W^* \delta$ is the scalar product, that is, it is the product of W^* and the component of δ measured in the direction of W^*. On the right-hand side, the summation extends over all the hinge discontinuities, and the integral extends over all the rest of the frame. The formation of the integral may be understood by noting that the rotation between two cross-sections distant ds apart, where the mean curvature is $\kappa = M/EI$, is simply $\kappa\, ds$, so that the virtual work done by the bending moment M^* is $M^* \kappa\, ds$.

Now a statically indeterminate frame is capable of supporting sets of self-stressing moments; it is possible for bending moments to exist in the frame which are in equilibrium with zero external load. Denoting such a set of bending moments by m, equation (5.3) becomes

$$\Sigma m\phi + \int m\frac{M}{EI} ds = 0. \tag{5.4}$$

If the number of redundancies in the complete elastic frame is denoted by R, then exactly R independent sets of self-stressing moments m can be constructed. Equation (5.4) thus represents a total of R independent equations, which suffice to determine the values of R unknown quantities. If the frame is entirely elastic, then all hinge discontinuities ϕ in equation (5.4) will be zero, and the R equations will yield the conventional elastic solution.

On the other hand, if the frame is collapsing by a regular mechanism with $(R+1)$ hinges, then all the bending moments in the frame will be known, but there will be $(R+1)$ unknown hinge rotations ϕ. The fact that there are only R equations reflects the fact that the collapse mechanism has one degree of freedom; the hinge rotations, which will involve terms due to the elastic curvatures of the members, can be found in terms of one of their number.

For a partial collapse mechanism, in which part of the frame is collapsing in a mechanism of one degree of freedom, and part remains indeterminate, there will again be $(R+1)$ unknown quantities, some of which are values of the redundancies while the rest are hinge rotations. Again it will be found that the redundant quantities can be calculated, and the hinge rotations determined to within one arbitrary quantity.

Once the state of the frame has been found by use of equation (5.4), in the sense that all bending moments (and hence curvatures) and the hinge rotations have been determined, the resulting deflexions may be computed. To find a deflexion Δ at a particular point in the structure, it is convenient to use again the virtual work equation with the *actual* deformations as the compatible set. If a unit load is applied at the particular point where Δ is to be computed, and *any* distribution of bending moments m^* constructed which is in equilibrium with that unit load, then equation (5.3) gives

$$1 \cdot \Delta = \Sigma m^* \phi + \int m^* \frac{M}{EI} ds. \tag{5.5}$$

All the quantities on the right-hand side are known, and the device of the unit load has picked out from the general left-hand side of equation (5.3) the precise deflexion Δ whose value is required.

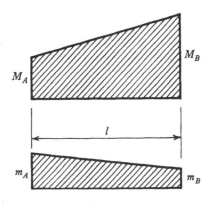

Fig. 5.2

Examples will make clear the use of equations (5.4) and (5.5). As an aid to the calculation of the integral, it may be noted that for a frame with straight members between joints, the residual moments m and the 'unit-load moments' m^* always vary linearly along the members. In addition, if point loads only act on the frame (for this purpose, any distributed load could be replaced by a series of point loads), the actual moments M also vary linearly between joints and loading points. If a span of length l has actual moments M which vary linearly between the values M_A and M_B, and the residual moments m vary between m_A and m_B, fig. 5.2, then

$$\int_0^l m \frac{M \, dx}{EI} = \frac{l}{6EI} [2(M_A m_A + M_B m_B) + (M_A m_B + M_B m_A)]. \tag{5.6}$$

(Similar standard integrals can of course be constructed for more complex patterns of loading.)

Before using these techniques to discuss the *collapse* state of a frame, it is of interest to trace the complete loading history of the fixed-ended beam whose load–deflexion curve was given as fig. 2.7 in vol. 1 (without, however, any analysis being presented in that volume); the load–deflexion curve is shown in fig. 5.6 later.

5.2 Fixed-ended beam

The fixed-ended beam of fig. 5.3 (a) is of uniform cross-section having flexural rigidity EI and full plastic moment M_p, and is acted upon by a single point load W; the value of W is increased slowly until collapse occurs. The most general bending moment diagram under this loading is shown in fig. 5.3 (b), drawn, as usual, with hogging bending moments positive; it is to be expected that the calculated value of M_B will be negative. Indeed, at collapse, the distribution of fig. 5.3 (c) shows hogging plastic hinges at the ends A and C and a sagging plastic hinge under the load.

The most general deformation pattern for the beam will involve elastic curvatures of the members between the critical sections A, B and C, with possible hinge discontinuities θ_A, θ_B and θ_C at the critical sections themselves, fig. 5.3 (d). These hinge discontinuities may have values of zero at any particular stage in the loading; when all three hinges have formed, collapse is of course occurring for this particular example. In any case, the value of θ_B may be expected to be zero or negative; the signs of the hinge rotations (again, as usual) must accord with the signs of the corresponding full plastic moments.

The fixed-ended beam has two redundancies, and a single relationship must therefore exist between the values of M_A, M_B and M_C in fig. 5.3 (b). Considering a straight bar virtual mechanism derived from fig. 5.3 (d), for which $\theta_A = \theta$, $\theta_C = 2\theta$, $\theta_B = -3\theta$, the equation of virtual work gives the equilibrium relation
$$M_A - 3M_B + 2M_C = 2Wl. \tag{5.7}$$

Two more equations can be found by consideration of distributions of self-stressing moments m. Figure 5.4 (a) shows the most general pattern of residual moment, specified in terms of two independent quantities m_A and m_C. Figure 5.4 (a) can be built up by the superimposition of arbitrary multiples of the two distributions of figs. 5.4 (b) and (c), and these two latter distributions will be used as the two independent distributions m

in equation (5.4). Since equation (5.4) is homogeneous, the scales of fig. 5.4(b) and (c) are immaterial.

Table 5.1 assembles the information required for the calculation of equation (5.4). Using the first and second lines of table 5.1, together with the hinge discontinuities of the last line, equation (5.4) gives

$$3\theta_A + \theta_B + \frac{2l}{6EI}[2(3M_A + M_B) + (M_A + 3M_B)] + \frac{l}{6EI}(2M_B + M_C) = 0,$$

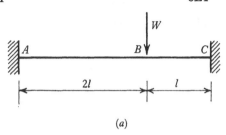

(a)

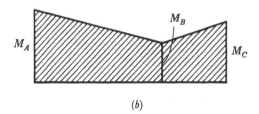

(b)

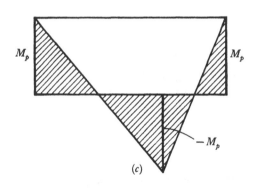

(c)

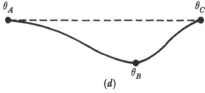

(d)

Fig. 5.3

95

that is, $\qquad 3\theta_A + \theta_B \qquad + \dfrac{l}{6EI}(14M_A + 12M_B + M_C) = 0.$ $\qquad$ (5.8)

Similarly, $\qquad 2\theta_B + 3\theta_C + \dfrac{l}{6EI}(4M_A + 15M_B + 8M_C) = 0$ $\qquad$ (5.9)

is the equation resulting from the use of the first and third lines of table 5.1. Equations (5.7), (5.8) and (5.9), together with the knowledge of the yield condition, furnish all the information necessary for determining the bending moments in the fixed-ended beam from first loading right up to collapse.

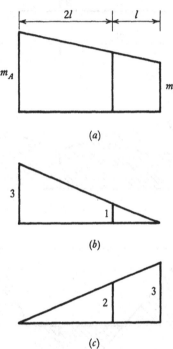

Fig. 5.4

Table 5.1

Distribution	Fig.	Span 1, 2l		Span 2, l	
		M_A	M_B	M_B	M_C
M	5.3 (b)	M_A	M_B	M_B	M_C
m_1	5.4 (b)	3	1	1	0
m_2	5.4 (c)	0	2	2	3
m^*	5.5 (c)	0	0	0	1
Discontinuities	5.3 (d)	θ_A	θ_B		θ_C

Initially, the beam will be entirely elastic; that is, all three of M_A, M_B and M_C will be numerically less than M_p. There will then be no hinge discontinuities, and θ_A, θ_B and θ_C in equations (5.8) and (5.9) must be set equal to zero. Equations (5.7), (5.8) and (5.9) may then be solved simultaneously to give the elastic solution:

$$M_A = \tfrac{6}{27}Wl, \quad M_B = -\tfrac{8}{27}Wl, \quad M_C = \tfrac{12}{27}Wl. \qquad (5.10)$$

This solution is valid so long as the largest bending moment, M_C, is less than M_p, that is, for $W < 9M_p/4l$. As soon as W exceeds this value, a plastic hinge will form at C, and some finite hinge discontinuity θ_C will develop. However, there are still only three unknown quantities in equations (5.7), (5.8) and (5.9), namely θ_C, M_A and M_B. The value of M_C is known to be equal to M_p, and, on the assumption that the numerical values of M_A and M_B are still less than M_p, the values of θ_A and θ_B continue to be zero.

Put another way, the deformation condition at C (zero slope at the fixed end) has been destroyed by the formation of a plastic hinge there, and so the condition is no longer available for use in the solution of the problem. However, this destruction of a deformation condition is exactly balanced by one extra piece of information; the value of the bending moment at the end in question is known to have the fully plastic value. Thus the three equations can again be solved to give

$$\frac{81}{28}\frac{M_p}{l} \geqslant W \geqslant \frac{9}{4}\frac{M_p}{l},$$

$$M_A = \tfrac{12}{27}Wl - \tfrac{1}{2}M_p, \quad M_B = -\tfrac{14}{27}Wl + \tfrac{1}{2}M_p, \quad M_C = M_p, \qquad (5.11)$$

$$\frac{EI\theta_C}{l} = \tfrac{1}{3}Wl - \tfrac{3}{4}M_p.$$

The value of θ_C must be positive, which confirms that $W \geqslant 9M_p/4l$. At $W = 81M_p/28l$, M_B becomes equal to $-M_p$, and a new plastic hinge forms; at this stage, the value of M_A is $\tfrac{11}{14}M_p$.

Thus, for loads whose values are increased still further, the continuity condition at B is destroyed, but again this is balanced by the knowledge of the value of M_B. Equation (5.7) may now be solved directly for the value of M_A, and substitution into equations (5.8) and (5.9) gives

$$3\frac{M_p}{l} > W \geqslant \frac{81}{28}\frac{M_p}{l},$$

$$M_A = 2Wl - 5M_p, \quad M_B = -M_p, \quad M_C = M_p, \qquad (5.12)$$

$$\frac{EI\theta_B}{l} = -\tfrac{14}{3}Wl + \tfrac{27}{2}M_p, \quad \frac{EI\theta_C}{l} = \tfrac{8}{3}Wl - \tfrac{15}{2}M_p.$$

At $Wl = 3M_p$, the value of M_A finally reaches M_p, and collapse ensues; this condition will be discussed more fully in section 5.3 later.

The load–deflexion curve corresponding to the results of equations (5.10), (5.11) and (5.12) may be calculated, as W is increased from zero up to the collapse value. If the deflexion at the loading point B is required, then, in accordance with equation (5.5), a bending moment distribution m^* must be constructed which is in equilibrium with a unit load placed

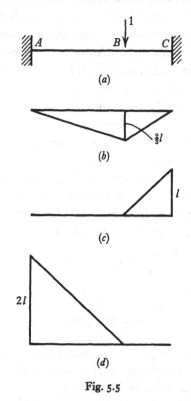

Fig. 5.5

at B, fig. 5.5 (a). There is an infinite number of such distributions, of which three simple ones are shown in fig. 5.5 (b), (c) and (d). The genesis of these particular three should be clear; fig. 5.5 (c), for example, corresponds to a unit load placed on a cantilever ABC in which the end A is free. It is this distribution which is entered in table 5.1. Thus, taking the first and fourth lines of table 5.1, equation (5.5) may be evaluated to give the deflexion Δ_W of the loading point:

$$\Delta_W = l\theta_C + \frac{l^2}{6EI}(2M_C + M_B). \tag{5.13}$$

The appropriate values of M_B, M_C and θ_C may be substituted into this expression to give the value of the deflexion.

In the elastic range, for example, for which $\theta_C = 0$, the values from equation (5.10) give, for $W \leqslant 9M_p/4l$,

$$\Delta_W = \frac{8}{81} \frac{Wl^3}{EI}. \tag{5.14}$$

Similarly, from equations (5.11) for $81M_p/28l \geqslant W \geqslant 9M_p/4l$,

$$\Delta_W = \frac{l^2}{EI}(\tfrac{20}{81}Wl - \tfrac{1}{3}M_p), \tag{5.15}$$

and, after the second hinge has formed, the use of equations (5.12) leads to

$$\Delta_W = \frac{l^2}{EI}(\tfrac{8}{3}Wl - \tfrac{22}{3}M_p). \tag{5.16}$$

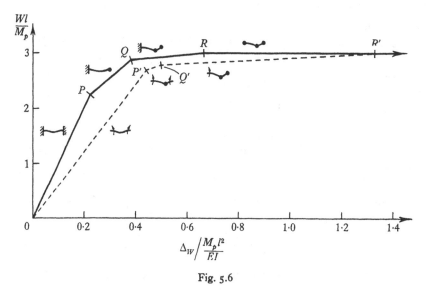

Fig. 5.6

These last three equations were plotted in vol. 1 (fig. 2.7), and are shown here by full lines in fig. 5.6; they give the complete load–deflexion curve up to the point of collapse.

Just at the point of incipient collapse, when $W = 3M_p/l$, the deflexion is given by $\Delta_W = 2M_p l^2/3EI$, from equation (5.16). As will be seen in section 5.3, it is not necessary to trace the whole history of loading in order to compute this maximum deflexion; the collapse state may be examined directly.

5. ELASTIC–PLASTIC ANALYSIS

In the calculations outlined above, it should be noted that certain conventional assumptions of elastic theory have been made. For example, it was assumed that the fixed-ended beam was initially stress free, that neither end settled with respect to the other, and that the slopes at the ends of the beam remained truly horizontal while the cross-sections there remained elastic. It is of interest to repeat the calculations assuming some more realistic (but still idealized) end conditions. If the supposedly fixed ends in fact rotate elastically under moment, there will exist a linear relation between the value M of the bending moment and the corresponding rotation θ; this relation will be taken, for simplicity of computation, as

$$\theta = \frac{Ml}{3EI}. \tag{5.17}$$

(Equation (5.17) does not represent an excessively flexible connexion, as may be seen from the resulting load–deflexion curve, shown by broken lines in fig. 5.6.)

For small values of W, for which the whole system is elastic, the elastic values of end rotation θ_A and θ_C may be introduced into equations (5.8) and (5.9) to give

$$\left.\begin{aligned} 20M_A + 12M_B + M_C &= 0, \\ 4M_A + 15M_B + 14M_C &= 0; \end{aligned}\right\} \tag{5.18}$$

the value of θ_B has been set equal to zero. Equations (5.18) may now be solved simultaneously with the invariant equilibrium equation (5.7), to give

$$M_A = 0.206Wl, \quad M_B = -0.372Wl, \quad M_C = 0.339Wl. \tag{5.19}$$

The corresponding elastic deflexion of the loading point is given, from equation (5.13), by

$$\Delta_W = \frac{l^2}{6EI}(4M_C + M_B) = 0.164\frac{Wl^3}{EI}. \tag{5.20}$$

The moment at end C has been relaxed so far that the first plastic hinge no longer forms at that section; instead, a hinge is formed at B at a load $W = 2.69M_p/l$, and the corresponding deflexion is $0.440M_p l^2/EI$ from equation (5.20). For increased loading the basic equations must be modified to include a finite hinge discontinuity θ_B at B, with $M_B = -M_p$; the work proceeds exactly as before, and the next hinge, at C, is found to form at $W = 2.79M_p/l$ and a deflexion $0.5M_p l^2/EI$. Final collapse occurs at a deflexion $4M_p l^2/3EI$, just double that of the corresponding perfectly encastré beam. The complete load–deflexion curve is shown by broken lines in fig. 5.6, and is compared with that of the 'perfect' beam.

This type of behaviour is typical of any practical structure, and must be borne in mind when making calculations involving deflexions; most such calculations assume an initially stress-free state. The deflexions of a real frame will be affected not only by flexible connexions, but also by any residual stress system that happens to be present before loading starts. Such residual stresses can alter the predicted order of formation of hinges, and may lead to either reduced or increased values of deflexions. Thus the calculations of the next section will be modified by the actual imperfections of a real frame.

5.3   Deflexions at collapse

Returning to the example of the 'perfect' fixed-ended beam, fig. 5.3, the final collapse equation, $3M_p = Wl$, is the limiting state of the equilibrium equation (5.7). The final collapse mechanism, fig. 5.3 (d), may be examined directly by means of equation (5.4), setting $M_A = -M_B = M_C = M_p$ at the hinge positions. In fact, equations (5.8) and (5.9) become

$$\left. \begin{aligned} 3\theta_A + \theta_B && +\frac{M_p l}{2EI} &= 0 \\[2mm] 2\theta_B + 3\theta_C &&-\frac{M_p l}{2EI} &= 0. \end{aligned} \right\} \tag{5.21}$$

and

These two equations solve to give

$$\left. \begin{aligned} \theta_A &= \theta_A, \\[2mm] \theta_B &= -3\theta_A - \frac{M_p l}{2EI}, \\[2mm] \theta_C &= 2\theta_A + \frac{M_p l}{2EI}. \end{aligned} \right\} \tag{5.22}$$

It may be noted that the simple addition of the three equations (5.22) gives $\theta_A + \theta_B + \theta_C = 0$; this is a geometrical constraint on the possible form of the general collapse mechanism, fig. 5.3 (d). In fact, the coefficients of θ_A in the expressions (5.22) represent the hinge rotations of the plastic collapse mechanism, fig. 5.7. The value of θ_A cannot be determined since the deformation of the collapse mechanism is of arbitrary magnitude; from the definition of collapse, deformation is possible without change in the value of the applied load W. Equations (5.22) represent therefore the hinge rotations of the collapse mechanism, on which are superimposed terms due to the elastic deformation of the members; from equation (5.13), the deflexion Δ in fig. 5.7 is given by

$$\Delta = l\theta_C + \frac{l^2}{6EI}(2M_p - M_p) = 2l\theta_A + \frac{2}{3}\frac{M_p l^2}{EI}. \tag{5.23}$$

Here again, the deflexion results from an elastic component superimposed on the deflexion resulting from a motion of the collapse mechanism.

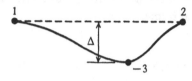

Fig. 5.7

At incipient collapse, point R in fig. 5.6, the hinge at A is just forming, and θ_A may be set equal to zero in equation (5.23) to give the deflexion at this state. Unless the complete elastic–plastic analysis has been made, however, it will not be known in general which of the hinges is last to form. Nevertheless, the hinge rotations at collapse of a frame can always be expressed in the form (5.22), in which one unknown quantity remains. It is then very easy to examine such expressions, and to determine the condition that one of the hinge rotations is zero, while all the others accord in sign with the full plastic moments acting at those hinges. Thus in equations (5.22) if either θ_B or θ_C were set equal to zero, on the assumption that one or other of these hinges was the last to form, the value of θ_A would be negative, while the full plastic moment at A has value $+M_p$. Thus, at the point of collapse, equations (5.22) reduce to

$$\theta_A = 0, \quad \theta_B = -\frac{M_p l}{2EI}, \quad \theta_C = \frac{M_p l}{2EI}. \tag{5.24}$$

When hinge rotations at incipient collapse have been found in this way, deflexions may be calculated from equation (5.5). In this connexion it may be noted that it is always possible to find at least one distribution m^*, in equilibrium with the unit load, whose values are zero at the hinge points where rotation has occurred. For this example, the distribution of fig. 5.5(d) has zero moments at B and C, and value $2l$ at A. To construct such a distribution in general, all that is necessary is to consider a frame similar to the original, but with frictionless pins replacing the rotating plastic hinges. Such a frame is still a structure, since it awaits the formation of the last hinge to turn it into a mechanism. A unit load acting on the new frame will generate automatically an equilibrium bending-moment distribution which has zero values at the hinges.

The deflexions at collapse of the portal frame of fig. 5.8(a) will now be computed; the frame has uniform section, full plastic moment M_p and flexural rigidity EI, has rigid connexions at the knees and perfectly encastré feet, and is initially stress free. The most interesting mode of collapse is that of fig. 5.8(b), in which one redundancy remains; the collapse relationships are $Vl = 8M_p$, $Hh \leqslant 2M_p$. The bending moment

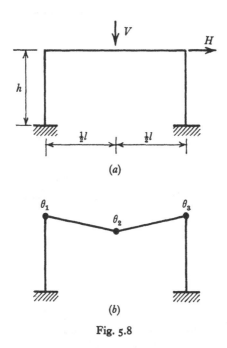

(a)

(b)

Fig. 5.8

distribution at collapse, M, is sketched in fig. 5.9(a); the bending moments at the feet of the columns have been denoted aM_p and bM_p. The usual sway mechanism may be used in the virtual work equation to give the equilibrium equation

$$Hh = (a-b)M_p. \tag{5.25}$$

Since the frame has three redundancies, three independent self-stressing distributions m may be constructed; those shown in fig. 5.9(b), (c) and (d) will be used, which are special cases of a more general pattern proposed below for the solution of multi-storey multi-bay frames (fig. 5.16). Table 5.2 may now be drawn up.

5. ELASTIC–PLASTIC ANALYSIS

The general deflected form of the frame is shown in fig. 5.8(b), and the hinge discontinuities in this figure have been entered as rotations (1) in table 5.2. Using the distribution m_1 from the table, equation (5.4) gives

$$\theta_1 + \tfrac{1}{2}\theta_2 + \frac{M_p h}{6EI}\left[3(a+1)\right] = 0. \qquad (5.26)$$

Table 5.2

Distribution	Fig.	Span 1, h		Span 2, $l/2$		Span 3, $l/2$		Span 4, h	
M	5.9(a)	aM_p	M_p	M_p	$-M_p$	$-M_p$	M_p	M_p	bM_p
m_1	5.9(b)	1	1	1	$\tfrac{1}{2}$	$\tfrac{1}{2}$	0	0	0
m_2	5.9(c)	0	0	0	$\tfrac{1}{2}$	$\tfrac{1}{2}$	1	1	1
m_3	5.9(d)	0	1	1	1	1	1	1	0
m_V^{*}	5.10(a)	$\tfrac{1}{2}l$	$\tfrac{1}{2}l$	$\tfrac{1}{2}l$	0	0	0	0	0
m_H^{*}	5.10(b)	h	0	0	0	0	0	0	0
Rotations (1) 5.8(b)		θ_1		θ_2		θ_3			
Rotations (2) 5.11		θ_1		θ_2		θ_3		θ_4	

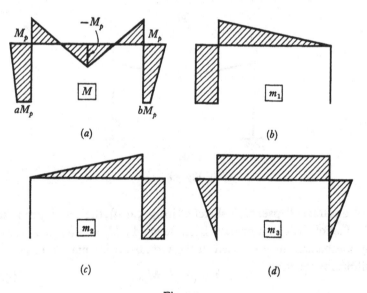

Fig. 5.9

Similarly, distributions m_2 and m_3 give

$$\tfrac{1}{2}\theta_2 + \theta_3 + \frac{M_p h}{6EI}\left[3(b+1)\right] = 0, \qquad (5.27)$$

and
$$\theta_1+\theta_2+\theta_3+\frac{M_p h}{6EI}[4+a+b]=0. \tag{5.28}$$

The hinge rotations θ may be eliminated from these three equations, to give
$$a+b=-1. \tag{5.29}$$

Such an elimination is always possible, leaving equations sufficient in number to determine the values of the redundancies. In this example, equation (5.29) may be solved simultaneously with equation (5.25) to give

$$\left.\begin{aligned}a &= -\tfrac{1}{2}+\frac{Hh}{2M_p},\\[1mm] b &= -\tfrac{1}{2}-\frac{Hh}{2M_p}.\end{aligned}\right\} \tag{5.30}$$

Thus the collapse bending-moment distribution, fig. 5.9(a) is now known uniquely, since the values a and b can be determined for any given value of the side load H.

For $H=0$, $a=b=-\tfrac{1}{2}$. As H is increased, the value of b approaches -1; indeed, at $Hh=M_p$, the moment at the right-hand column foot reaches the value $-M_p$, while a is zero. The problem must be discussed in two stages, since evidently the present analysis will hold only for $Hh \leqslant M_p$.

Case (a). $Vl=8M_p$, $Hh \leqslant M_p$.

Using the values of a and b from (5.30), equations (5.26) and (5.27) may be rearranged to give

$$\left.\begin{aligned}\theta_2 &= -2\theta_1-\frac{M_p h}{6EI}\left[3\left(1+\frac{Hh}{M_p}\right)\right],\\[1mm] \theta_3 &= \theta_1+\frac{M_p h}{6EI}\left[3\left(\frac{Hh}{M_p}\right)\right].\end{aligned}\right\} \tag{5.31}$$

From the condition that the signs of hinge rotations and full plastic moments must be the same for any particular hinge, it is seen that the hinge rotations at incipient collapse are

$$\left.\begin{aligned}\theta_1 &= 0,\\[1mm] \theta_2 &= -\frac{M_p h}{6EI}\left[3\left(1+\frac{Hh}{M_p}\right)\right],\\[1mm] \theta_3 &= \frac{M_p h}{6EI}\left[3\left(\frac{Hh}{M_p}\right)\right].\end{aligned}\right\} \tag{5.32}$$

For the purpose of calculating deflexions at the loading points, the two unit-load equilibrium distributions shown in fig. 5.10 will be used; these have been entered in table 5.2 as m_V^* and m_H^*. (Both distributions involve zero bending moments at the sections where there are hinge discontinuities θ_2 and θ_3.) Using equation (5.5):

$$\Delta_V = \frac{M_p h}{6EI}\left[\frac{3l}{2}(a+1)\right] + \frac{M_p l}{12EI}\left(\frac{l}{2}\right), \tag{5.33}$$

which gives, on substituting the value of a from (5.30),

$$\Delta_V = \frac{M_p l^2}{24EI}\left[1 + \frac{3h}{l}\left(1 + \frac{Hh}{M_p}\right)\right]. \tag{5.34}$$

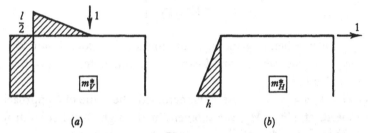

Fig. 5.10

Fig. 5.11

Similarly, the use of the m_H^* distribution in table 5.2 gives

$$\Delta_H = \frac{M_p h^2}{6EI}\left(\frac{Hh}{M_p}\right). \tag{5.35}$$

Case (b). $Vl = 8M_p$, $M_p \leqslant Hh \leqslant 2M_p$.

It was seen that for $Hh = M_p$ a plastic hinge developed at the foot of the right-hand column. For greater values of H, the analysis will proceed on the assumption (which can be checked from the final results, see equations (5.38)) that a plastic hinge with rotation θ_4 develops at the column foot, fig. 5.11; these hinge rotations are entered in the last line of table 5.2.

Once again, although a new unknown θ_4 has been introduced into the problem, the value of the bending moment at the right-hand column foot is now known to be equal to $-M_p$. (Thus the calculated value of θ_4 should also be negative.) Setting $b = -1$, equation (5.25) gives

$$a = -1 + \frac{Hh}{M_p}. \tag{5.36}$$

Equations (5.26), (5.27) and (5.28) are replaced by

$$
\left.
\begin{aligned}
\theta_1 + \tfrac{1}{2}\theta_2 \qquad\quad + \frac{M_p h}{6EI}\left(3\frac{Hh}{M_p}\right) &= 0, \\
\tfrac{1}{2}\theta_2 + \theta_3 + \theta_4 \qquad\qquad\qquad\ &= 0, \\
\theta_1 + \theta_2 + \theta_3 \quad + \frac{M_p h}{6EI}\left(2 + \frac{Hh}{M_p}\right) &= 0.
\end{aligned}
\right\} \tag{5.37}
$$

These equations may be solved in terms of θ_1 to give

$$
\left.
\begin{aligned}
\theta_2 &= -2\theta_1 - \frac{M_p h}{6EI}\left[6\frac{Hh}{M_p}\right], \\
\theta_3 &= \quad \theta_1 + \frac{M_p h}{6EI}\left[5\frac{Hh}{M_p} - 2\right], \\
\theta_4 &= \qquad\quad \frac{M_p h}{6EI}\left[2 - \frac{2Hh}{M_p}\right].
\end{aligned}
\right\} \tag{5.38}
$$

It will be seen that θ_4 is negative only if $Hh > M_p$, which confirms the range of validity of this analysis. The absence of a term in θ_1 in the expression for θ_4 indicates that no *plastic* motion of the final collapse mechanism is possible which involves a rotation at the right-hand column foot.

The last hinge to form is again at the left-hand end of the beam, so that the hinge rotations at incipient collapse may be found from (5.38) by setting $\theta_1 = 0$. Using again the m^* distributions of fig. 5.10, the deflexions of the loading points at incipient collapse are

$$
\left.
\begin{aligned}
\Delta_V &= \frac{M_p l^2}{24EI}\left[1 + \frac{h}{l}\left(6\frac{Hh}{M_p}\right)\right], \\
\Delta_H &= \frac{M_p h^2}{6EI}\left[2\frac{Hh}{M_p} - 1\right].
\end{aligned}
\right\} \tag{5.39}
$$

The deflexions at incipient collapse given by equations (5.34), (5.35) and (5.39) are displayed in fig. 5.12.

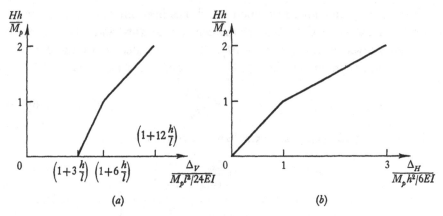

Fig. 5.12

5.4 A four-storey frame

The four-storey frame of example 4.14 in vol. 1 has been examined repeatedly in the literature of plastic theory; the frame is redrawn in fig. 5.13. The two columns are of the same section throughout, full plastic moment 355 kN m; no allowance will be made in the present example for the reduction in the value of full plastic moment due to axial load. The two upper beams have full plastic moment 531 kN m, and the two lower 583 kN m; all connexions are full strength, and the feet of the columns are fixed. The loads shown, measured in kN, must be multiplied by a factor 2.23 to produce the collapse mechanism of fig. 5.14; this mechanism may be found most easily by using the method of combination of mechanisms.

It will be seen that there are ten hinges in fig. 5.14, and the frame has twelve redundancies; it is to be expected, therefore, that three redundancies will remain. In fig. 5.15 is shown the bending-moment distribution at collapse, in which three unknown quantities M_a, M_b and M_c have been introduced; only the bending moments in the columns are statically indeterminate, and it will be seen that these bending moments fall into convenient groups. Using the fact that the maximum permitted bending moment in the columns is 355 kN m, the equilibrium distribution of fig. 5.15 will satisfy the yield condition for values of the unknown bending moments lying in the ranges

$$\left.\begin{aligned} -176 &\geqslant M_a \geqslant -355, \\ -228 &\geqslant M_b \geqslant -355, \\ -228 &\geqslant M_c \geqslant -355. \end{aligned}\right\} \qquad (5.40)$$

Thus the collapse mode of fig. 5.14 is confirmed, since it is certainly *possible* to satisfy the yield condition. The *actual* values of the three moments can be found by writing equation (5.4) twelve times (corresponding to the twelve redundancies of the frame), and solving these equations for the three unknown bending moments and the ten unknown

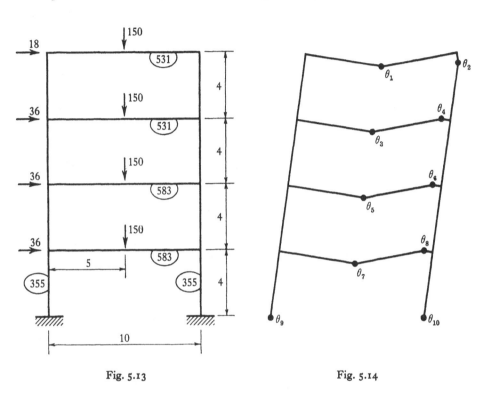

Fig. 5.13   Fig. 5.14

hinge rotations of fig. 5.14 (as usual, the hinge rotations can be expressed in terms of one of their number). The general patterns of residual moment sketched in fig. 5.16 will be found convenient for the analysis of larger multi-storey multi-bay frames, although other patterns can of course be constructed. The three distributions are applied to each beam in turn, and will provide $3MN$ equations for a plane frame having M storeys and N bays.

The moments of inertia of the upper beams, lower beams and columns will be taken as 370, 410 and $143 \times 10^{-6}\,\mathrm{m^4}$ respectively;

$$E = 205 \times 10^6\,\mathrm{kN/m^2}.$$

With these values, the residual patterns of fig. 5.16, used in conjunction with the collapse bending moments of fig. 5.15 and the corresponding hinge rotations of fig. 5.14, lead to the following twelve equations:

$$
\left.
\begin{aligned}
\tfrac{1}{2}\theta_1 \qquad\qquad +\theta_9 \qquad\qquad +(\tfrac{1}{14700})\,[2M_a+2M_b+2M_c+1223] &= 0,\\
\tfrac{1}{2}\theta_1+\theta_2 \qquad\qquad +\theta_{10}(+\theta_{11})+(\tfrac{1}{14700})\,[2M_a+2M_b+2M_c+1604] &= 0,\\
\theta_1+\theta_2 \qquad\qquad\qquad\qquad +(\tfrac{1}{14700})\,[\tfrac{8}{3}M_a \qquad\qquad\qquad + 212] &= 0,\\
\tfrac{1}{2}\theta_3 \qquad\qquad +\theta_9 \qquad\qquad +(\tfrac{1}{14700})\,[\;M_a+2M_b+2M_c+ 846] &= 0,\\
\tfrac{1}{2}\theta_3+\theta_4 \qquad +\theta_{10}(+\theta_{11})+(\tfrac{1}{14700})\,[\;M_a+2M_b+2M_c+1306] &= 0,\\
\theta_3+\theta_4 \qquad\qquad\qquad +(\tfrac{1}{14700})\,[\tfrac{8}{3}M_a+\tfrac{8}{3}M_b \qquad\quad + 265] &= 0,\\
\tfrac{1}{2}\theta_5 \qquad\qquad +\theta_9 \qquad\qquad +(\tfrac{1}{14700})\,[\qquad\quad M_b+2M_c+ 546] &= 0,\\
\tfrac{1}{2}\theta_5+\theta_6 \qquad +\theta_{10}(+\theta_{11})+(\tfrac{1}{14700})\,[\qquad\quad M_b+2M_c+ 763] &= 0,\\
\theta_5+\theta_6 \qquad\qquad (+\theta_{11})+(\tfrac{1}{14700})\,[\qquad\tfrac{4}{3}M_b+\tfrac{8}{3}M_c+ 197] &= 0,\\
\tfrac{1}{2}\theta_7 \quad +\theta_9 \qquad\qquad\qquad +(\tfrac{1}{14700})\,[\qquad\qquad\qquad M_c+ 283] &= 0,\\
\tfrac{1}{2}\theta_7+\theta_8 \quad +\theta_{10} \qquad\qquad +(\tfrac{1}{14700})\,[\qquad\qquad\qquad M_c+ 180] &= 0,\\
\theta_7+\theta_8 \qquad\qquad\qquad +(\tfrac{1}{14700})\,[\qquad\qquad\quad \tfrac{4}{3}M_c+ 214] &= 0.
\end{aligned}
\right\}
$$

$$(5.41)$$

In these equations, the bracketed term $(+\theta_{11})$ should be ignored for the moment.

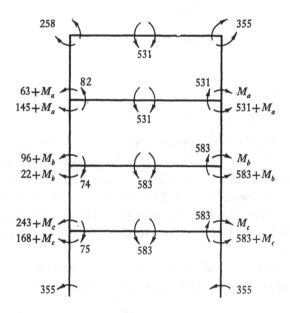

Fig. 5.15

Taking equations (5.41) in groups of three,

$$\theta_9 + \theta_{10}(+\theta_{11}) + (\tfrac{1}{14700})[\tfrac{10}{3}M_a + 4M_b + 4M_c + 2615] = 0,$$
$$\theta_9 + \theta_{10}(+\theta_{11}) + (\tfrac{1}{14700})[\tfrac{2}{3}M_a + \tfrac{10}{3}M_b + 4M_c + 1886] = 0,$$
$$\theta_9 + \theta_{10} \qquad + (\tfrac{1}{14700})[\qquad \tfrac{2}{3}M_b + \tfrac{10}{3}M_c + 1113] = 0,$$
$$\theta_9 + \theta_{10} \qquad + (\tfrac{1}{14700})[\qquad\qquad\qquad \tfrac{2}{3}M_c + 250] = 0.$$

$$(5.42)$$

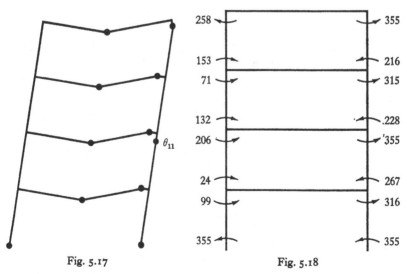

(a)   (b)   (c)

Fig. 5.16

Still ignoring (θ_{11}), it will be seen that $(\theta_9 + \theta_{10})$ can be eliminated, leaving three equations for M_a, M_b and M_c; these equations solve to give $M_a = -233$, $M_b = -161$, $M_c = -283\,\mathrm{kN\,m}$. Comparing these values with (5.40), it will be seen that the value $M_b = -161$ is not possible; a plastic hinge must form at the cross-section where an extra hinge rotation θ_{11} is marked in fig. 5.17. The formation of this hinge fixes the value of M_b at $-228\,\mathrm{kN\,m}$.

θ_{11}

258   355

153   216
71   315

132   .228
206   '355

24   267
99   316

355   355

Fig. 5.17   Fig. 5.18

A new analysis must be made, exactly as for the single-bay frame of fig. 5.11. However, equations (5.41) and (5.42) are unchanged except for the addition of the terms (θ_{11}) and for the fact that M_b now has the known value of -228. Solving the equations as before, it is found that

$$M_a = -216 \quad \text{and} \quad M_c = -267,$$

and the final bending moments for the columns are displayed in fig. 5.18. The complete set of equations may now be solved in terms of one hinge discontinuity, say the rotation θ_9 of the left-hand column foot, fig. 5.14, to give

$$
\begin{array}{rcll}
\theta_1 = & -2\theta_9 + 398(\tfrac{1}{14700}) & = & -322(\tfrac{1}{14700}) \\
\theta_2 = & 2\theta_9 - 466(\tfrac{1}{14700}) & = & 254(\tfrac{1}{14700}) \\
\theta_3 = & -2\theta_9 + 720(\tfrac{1}{14700}) & = & 0 \\
\theta_4 = & 2\theta_9 - 545(\tfrac{1}{14700}) & = & 175(\tfrac{1}{14700}) \\
\theta_5 = & -2\theta_9 + 430(\tfrac{1}{14700}) & = & -290(\tfrac{1}{14700}) \\
\theta_6 = & 2\theta_9 - 302(\tfrac{1}{14700}) & = & 418(\tfrac{1}{14700}) \\
\theta_7 = & -2\theta_9 - 32(\tfrac{1}{14700}) & = & -752(\tfrac{1}{14700}) \\
\theta_8 = & 2\theta_9 + 174(\tfrac{1}{14700}) & = & 894(\tfrac{1}{14700}) \\
\theta_9 = & \theta_9 & = & 360(\tfrac{1}{14700}) \\
\theta_{10} = & -\theta_9 - 72(\tfrac{1}{14700}) & = & -432(\tfrac{1}{14700}) \\
\theta_{11} = & 157(\tfrac{1}{14700}) & = & 157(\tfrac{1}{14700})
\end{array}
\qquad (5.43)
$$

All these hinge rotations must accord in sign with the full plastic moments; that is, all signs must be the same as the coefficients of θ_9 which represent to some scale the hinge rotations of the final collapse mechanism (it is known that θ_9 itself is positive). For $\theta_9 = 360(\tfrac{1}{14700})$, the value of θ_3 is just zero, and the final hinge rotations at the point of collapse are the figures on the right-hand sides of equations (5.43).

The known bending moments and hinge rotations may now be used to compute deflexions at incipient collapse of the frame. For example, the sidesway deflexion at the top of the frame may be computed by means of equation (5.5) by using the m^* distribution of fig. 5.19. The sway Δ_s is given by

$$\Delta_s = 8[-\theta_{11} - 2\theta_{10} + \tfrac{1}{14700}\{M_a + 2M_b + 3M_c + 1414\}]. \quad (5.44)$$

Inserting the values $M_a = -216$, $M_b = -228$ and $M_c = -267$ kN m, and taking the values of θ_{10} and θ_{11} from (5.43), the sway deflexion is determined as

$$\Delta_s = 8(\tfrac{649}{14700}) = 0.355 \text{ m.} \quad (5.45)$$

This sway *at collapse* might well be of acceptable magnitude; it must be emphasized that the calculations have been made for an unclad framework, assumed to be initially stress free and to have perfect connexions.

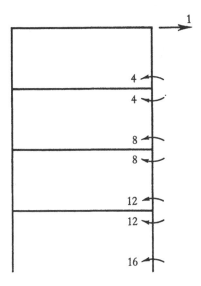

Fig. 5.19

5.5 The combination of mechanisms

The construction of a statically admissible bending-moment distribution for the collapsing frame of fig. 5.14 was relatively easy, involving only three unknown bending moments, M_a, M_b and M_c (fig. 5.15). It was shown that if inequalities (5.40) were satisfied (as they can be) then the mode of collapse was correct. For more complex frames, or for similar frames where fewer hinges are formed at collapse, the construction is not so easy, and the establishment of bounds on the permitted values of unknown bending moments may be prohibitively difficult for manual calculation. On the other hand, elastic–plastic calculations like those of section 5.4 above can also be very lengthy, particularly if the number of bending moments to be determined is large. It is, however, possible to extend the method of combination of mechanisms so that a statically admissible bending-moment distribution can be derived without the necessity of a statical analysis.

As a simple numerical example, the rectangular portal frame of fig. 5.20 collapses in the beam mode at a load factor $\lambda = 2\cdot4$. At collapse, the bending moments will be as shown in fig. 5.21, where the value of M is not determined by the plastic analysis. By inspection, it will be seen that any value of M in the range $-1\cdot2 \leqslant M \leqslant 6$ will enable the yield condition to be satisfied, so that the collapse mechanism is confirmed. (The elastic–plastic analysis of section 5.3 gives $M = 5.4$, as may be checked from the second of equations (5.30), from which $b = -0\cdot9$; see fig. 5.31 (c) later.)

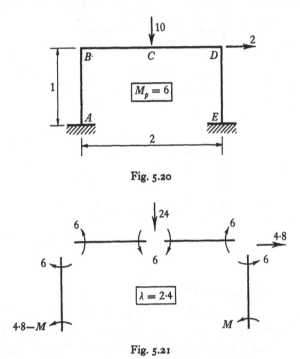

Fig. 5.20

Fig. 5.21

In extending the method of combination of mechanisms to derive automatically values of the unknown bending moments which satisfy the yield condition, the work equation will be used in the form

$$\lambda\Sigma(M_F)_i\,\phi_i = \Sigma(M_p)_i\,|\phi_i|. \tag{5.46}$$

In this equation, the ϕ_i are hinge rotations of any mechanism, and $(M_p)_i$ are the values of the full plastic moments at the corresponding sections i. The values $(M_F)_i$ correspond to *any* set of bending moments in equilibrium with the applied loads under unit load factor. Thus, for example, the frame of fig. 5.20 may be made statically determinate by the insertion of three pins, fig. 5.22. A possible distribution M_F would then be

$$(M_F)_i \equiv [2, 0, -5, 0, 0].\tag{5.47}$$

The three possible modes of collapse are shown in fig. 5.23, and, using the uniform value $M_p = 6$, equation (5.46) gives

$$(a) \quad 2\lambda = 4(6), \quad \lambda = 12.0;$$
$$(b) \quad 10\lambda = 4(6), \quad \lambda = 2.4; \qquad (5.48)$$
$$(c) \quad 12\lambda = 6(6), \quad \lambda = 3.0;$$

These calculations confirm mode (b) to be correct, since the value of λ is least.

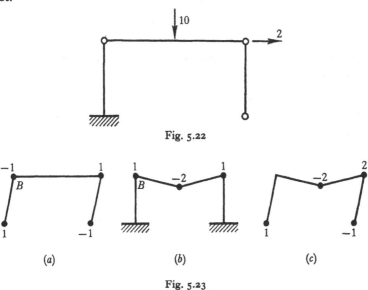

Fig. 5.22

(a)   (b)   (c)

Fig. 5.23

Now a partial collapse mechanism of this sort is in some sense wasteful; it seems likely that an economical frame would collapse by the formation of an *over-complete* mechanism rather than in a mode which has too few hinges to make the frame statically determinate. (These intuitive ideas are confirmed by the minimum-weight analysis of chapter 7.) With this in mind, the full plastic moments of the critical sections involved in the partial collapse mechanism will be supposed to be increased, so that the collapse mechanism is inhibited. The loads on the frame could then be increased, that is, λ can be increased above the value 2.4, until the frame reaches a new critical state when one or more extra hinges are formed, permitting a mode of collapse by some other mechanism.

Accordingly, the full plastic moments at the knees of the frame of fig. 5.20, and at the centre of the beam, will be supposed to have the value

$2 \cdot 5\lambda$ instead of 6. Thus, as λ is increased from the collapse value of $2 \cdot 4$, the full plastic moments at these three sections increase exactly in proportion. The full plastic moments for the frame as a whole are now

$$(M_p)_i \equiv [6, 2 \cdot 5\lambda, 2 \cdot 5\lambda, 2 \cdot 5\lambda, 6], \qquad (5.49)$$

and these will be used with the set (5.47) for $(M_F)_i$ in equation (5.46). For mechanisms (a) and (b) of fig. 5.23,

$$\begin{aligned}(a) & \quad 2\lambda = 12 + 5\lambda, \\ (b) & \quad 10\lambda = 10\lambda,\end{aligned} \qquad (5.50)$$

and these equations may be compared with the first two of (5.48). It will be seen that mechanism (b) is just formed at any value of λ, while mechanism (a) is impossible in that a negative value of λ results. Indeed, no meaningful value of λ can be calculated from either of equations (5.50). Combining the mechanisms in the usual way, however, and subtracting 5λ from the right-hand side to allow for the suppression of the hinge at B, equations (5.50) lead to

$$(c) \quad 12\lambda = 12 + 10\lambda; \quad \lambda = 6 \cdot 0. \qquad (5.51)$$

Thus the new, artificially strengthened frame collapses at a load factor of 6.

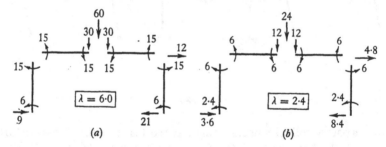

Fig. 5.24

In this new collapse mode there exist the four hinges of fig. 5.23 (c); in addition, by the device of strengthening the previously formed hinges in proportion to the load factor, the three hinges of fig. 4.23 (b) have been retained, two of which are common with hinges of fig. 5.23 (c). Thus there are now five plastic hinges in the artificially strengthened frame, and the bending-moment diagram can be determined uniquely. (The reason for the existence of *five* rather than four hinges is given below.)

Figure 5.24 (a) gives the statical analysis at a load factor $\lambda = 6$; the values of the full plastic moments at the hinges can be written immediately from set (5.49). If fig. 5.24 (a) is now redrawn with every numerical value reduced in the ratio 6·0/2·4, then fig. 5.24(b) results. Comparing figs. 5.21 and 5.24(b), it is evident that the technique just outlined has generated a set of statically admissible bending moments for the true collapse state of the frame.

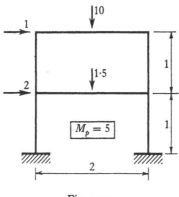

Fig. 5.25

The two-storey frame of fig. 5.25 involves more work in the analysis. The four basic independent mechanisms are shown in fig. 5.26; if to these mechanisms are added elementary joint rotations, all possible collapse mechanisms for the whole frame may be derived. Corresponding to fig. 5.26,

$$
\begin{aligned}
&(a) \quad \lambda = 20, \quad \lambda = 20\cdot00; \\
&(b) \quad 3\lambda = 20, \quad \lambda = 6\cdot67; \\
&(c) \quad 10\lambda = 20, \quad \lambda = 2\cdot00; \\
&(d) \quad 1\cdot5\lambda = 20, \quad \lambda = 13\cdot33.
\end{aligned}
\qquad (5.52)
$$

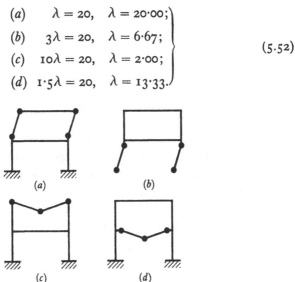

Fig. 5.26

There is in fact no combination of these mechanisms which appears to give a value of the load factor less than 2·00, and mechanism (c), which leaves the frame four times redundant, must be assumed tentatively to be the correct collapse mechanism.

Following the technique proposed, the artificially strengthened frame of fig. 5.27(a) will be examined, in which the full plastic moment of the

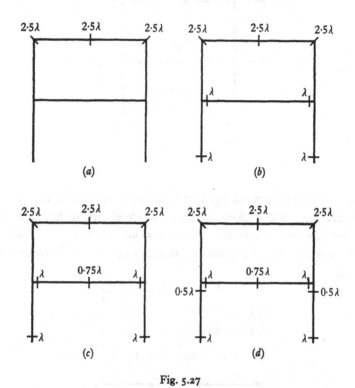

Fig. 5.27

three critical sections marked has been given the value 2·5λ, the full plastic moment at all other critical sections remaining at the value $M_p = 5$. The four independent equations now become

$$
\begin{aligned}
(a) \quad & \lambda = 10 + 5\lambda, \\
(b) \quad & 3\lambda = 20, \quad \lambda = 6\cdot67; \\
(c) \quad & 10\lambda = 10\lambda, \\
(d) \quad & 1\cdot5\lambda = 20, \quad \lambda = 13\cdot33.
\end{aligned}
\right\} \tag{5.53}
$$

Combining mechanisms in the usual way (cf. section 4.3 of vol. 1), and following the sequence of fig. 5.28,

$$
\begin{array}{lll}
(a) & \lambda = 10 + 5\lambda \\
(b) & 3\lambda = 20 \\
\hline
& 4\lambda \quad 30 + 5\lambda
\end{array}
$$

Rotate joints:
$$10$$

$$
\begin{array}{lll}
(e) & 4\lambda = 20 \ +5\lambda \\
(c) & 10\lambda = \quad 10\lambda \\
\hline
& 14\lambda \quad 20 + 15\lambda
\end{array}
$$

Cancel hinge:
$$5\lambda$$

$$
\begin{array}{lll}
(f) & 14\lambda = 20 + 10\lambda; \quad \lambda = 5\text{·}00, \\
(d) & 1\text{·}5\lambda = 20 \\
\hline
& 15\text{·}5\lambda \quad 40 + 10\lambda
\end{array}
$$

Cancel hinge:
$$10$$

$$(g) \quad 15\text{·}5\lambda = 30 + 10\lambda; \quad \lambda = 5\text{·}45.$$

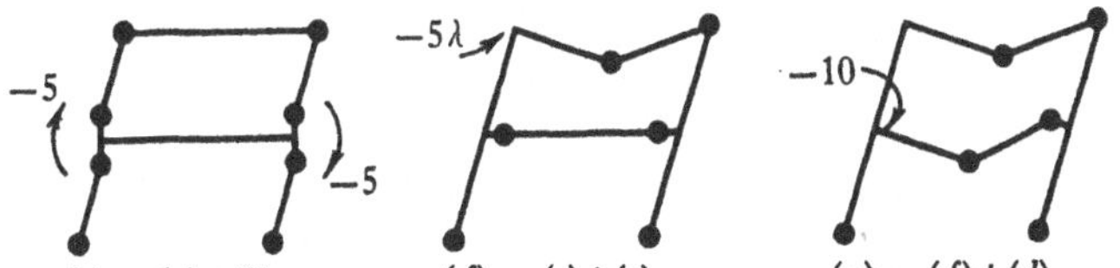

$(e) = (a)+(b)$ $(f) = (e)+(c)$ $(g) = (f)+(d)$

Fig. 5.28

It seems that mechanism (f) of fig. 5.28 is now the correct collapse mechanism for the strengthened frame of fig. 5.27(a). Accordingly, the frame of fig. 5.27(b) will now be analysed, in which the newly formed hinges are assigned full plastic moments $1\text{·}00\lambda$ (since the newly found load factor was $\lambda = 5\text{·}00$, and originally $M_p = 5$). Although fig. 5.27(b) has seven hinges, a static check at $\lambda = 5\text{·}00$ reveals that it is still once indeterminate (for a reason to be given below), and the work must proceed as before. The four basic equations for the frame of fig. 5.27(b) are

$$
\left.
\begin{array}{lll}
(a) & \lambda = 10 + 5\lambda, \\
(b) & 3\lambda = 10 + 2\lambda, \quad \lambda = 10\text{·}00; \\
(c) & 10\lambda = \quad 10\lambda, \\
(d) & 1\text{·}5\lambda = 10 + 2\lambda.
\end{array}
\right\} \quad (5.54)
$$

These equations can be combined again in the sequence illustrated in fig. 5.28; alternatively, a start may be made from mechanism (f), the previous collapse mechanism, for which may be written immediately:

$$(f) \quad 14\lambda = \quad 14\lambda.$$

Now adding:
$$(d) \quad 1\cdot5\lambda = 10+2\lambda$$

$$\overline{15\cdot5\lambda \quad 10+16\lambda}$$

Cancel hinge:
$$\overline{ \quad 2\lambda}$$

$$(g) \quad 15\cdot5\lambda = 10+14\lambda; \quad \lambda = 6\cdot67.$$

No other combination appears to give a lower value of λ, and the next move must be to analyse the frame of fig. 5.27(c), which is still once indeterminate at $\lambda = 6\cdot67$; the single newly formed hinge is given a full plastic moment of $0\cdot75\lambda$. (Note that $(0\cdot75)(6\cdot67) = 5$.) The four basic equations become

$$
\left.
\begin{aligned}
(a) & \quad \lambda = 10+5\lambda, \\
(b) & \quad 3\lambda = 10+2\lambda, \quad \lambda = 10\cdot00; \\
(c) & \quad 10\lambda = \quad 10\lambda, \\
(d) & \quad 1\cdot5\lambda = \quad 3\cdot5\lambda,
\end{aligned}
\right\} \tag{5.55}
$$

and it appears that no combination of these equations will lead to a value of λ less than $10\cdot00$. Accordingly, the values of the full plastic moments at the two newly formed hinges (corresponding to mechanism (b) of fig. 5.26) will be maintained at $0\cdot5\lambda$, fig. 5.27(d). At this stage, however, the frame is at last statically determinate, and a complete bending-moment distribution may be constructed.

This distribution is shown in fig. 5.29, drawn directly for $\lambda = 2$ and using the values noted in fig. 5.27(d). It will be seen that the distribution is one which corresponds to the original collapse of the upper beam, and which everywhere satisfies the yield condition.

The frame of fig. 5.25 has six redundancies, and seven hinges should therefore be required for the formation of a regular collapse mechanism. At the end of the first analysis, however, only three hinges had been formed, leaving four redundancies. By artificially strengthening the frame, as in fig. 5.27(a), an extra condition is imposed; that is, the full plastic moment at the three critical sections of the top beam is required to be maintained at a certain value $(2\cdot5\lambda)$. Effectively, therefore, this artificially strengthened frame requires the formation of one further hinge if it is to become statically determinate; the frame of fig. 5.27(a) should collapse therefore with eight hinges. In fact, fig. 5.27(*b*) displays only seven hinges, so that one indeterminacy is left. The frame of fig. 5.27(*b*) should in turn collapse with nine hinges since, again, one further condition is imposed; however, only eight hinges were formed, fig. 5.27(*c*). This last frame should then require ten hinges for statical determinacy, and this was indeed finally achieved in fig. 5.27(*d*).

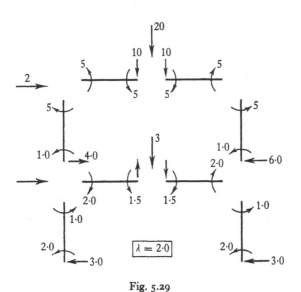

Fig. 5.29

The four-storey frame of fig. 5.13 may now be re-examined. The first artificially strengthened frame, corresponding to the collapse mechanism of fig. 5.14, $\lambda = 2\cdot23$, is that shown in fig. 5.30(*a*). Upon analysis, three steps are required to make the frame statically determinate (two of these steps are simultaneous, in that two separate mechanisms are formed together), bringing the total number of hinges up to sixteen. The final state is shown in fig. 5.30(*b*), and substitution of the collapse value $\lambda = 2\cdot23$ confirms that a statically admissible state can be found for the collapse mechanism of fig. 5.14. The extra hinges formed in fig. 5.30(*b*) all result from elementary joint rotations, and this type of mechanism has to be considered when analysing the artificial frames.

The numerical values shown in fig. 5.30(*b*) make it clear that the technique described here gives some sort of 'averaging-out' process; the bending moments at the right-hand ends of the beams, for example, are split equally between upper and lower columns, and the two moments at the column feet are equalized. Similarly, the simple frame of fig. 5.24(b) had equal moments at the feet. This particular frame is discussed briefly in the next section.

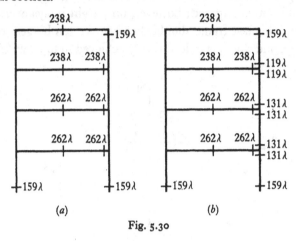

Fig. 5.30

5.6 The design of columns; preliminary remarks

More general discussion of the problem of column design will be given in chapter 9. However, the results for the simple rectangular frame of fig. 5.20(a) may be collected here as an illustration of the difficulties facing the designer. At collapse, the statical analysis of fig. 5.21 is valid, as was seen, for $-1 \cdot 2 \leqslant M \leqslant 6$. The two extreme cases are illustrated in fig. 5.31(a) and (b). That of fig. 5.31(a) seems implausible; the moment $(+1 \cdot 2)$ at the foot of the right-hand column seems to be of the wrong sign for a horizontal load acting from left to right. However, this is almost certainly the worst distribution, since the left-hand column is, correspondingly, bent in single curvature, which is the most critical case for column design.

Figure 5.31(b) appears to give a more likely distribution, and this is supported by the elastic–plastic analysis of fig. 5.31(c). The distribution of fig. 5.24(b), redrawn in fig. 5.31(d), which was generated by considering an artificially strengthened frame, is also not dissimilar. For comparison, fig. 5.31(e) shows the fully elastic solution for $\lambda = 1$ scaled up by the collapse factor of 2.4, with no notice taken of the fact that the full plastic moment $(M_p = 6)$ has been (hypothetically) exceeded. It will be argued later that this last distribution might well be used for column design, but in fact in this simple example only the distribution of fig. 5.31(a) stands out as being markedly different from the others.

It was said that fig. 5.31(*a*) seemed implausible, but this judgment is based on preconceived notions of elastic perfection. If the frame had been assembled with some considerable prestress, then the distribution of fig. 5.31(*a*) could result after loading to collapse, even if the assumptions of rigid joints, zero settlement, and so on, had been obeyed. Alternatively, settlement of the left-hand column foot could modify the elastic–plastic distribution of fig. 5.31(*c*) a good way towards that of fig. 5.31(*a*).

The designer will have to decide which of the five distributions in fig. 5.31 seems to him to be the most *reasonable* for design, bearing in mind known (or unknown) conditions of fabrication, erection, and the state of the foundations. There is no need to take the worst of the five, if this is considered to be an unlikely distribution for a given building on a given site, but at least fig. 5.31(*a*) and (*b*) give limits between which the column moments must fall. Fortunately, column design is often not a critical part of the process of the design of the frame as a whole; that is, a conservative estimate of conditions in the columns will lead to only a small penalty in weight and cost.

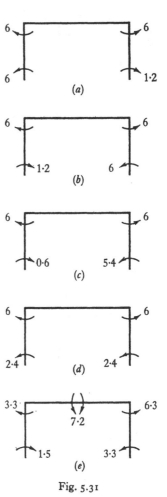

Fig. 5.31

These ideas will be considered further in chapter 9; they will be set on one side for the moment while other aspects of frame design are discussed.

EXAMPLES

5.1 The bending-moment distribution of fig. 5.5(*c*) is entered as the set m^* in table 5.1, and was used to derive equation (5.13) for the deflexion Δ_W at the loading point of the fixed-ended beam. Confirm that equation (5.13) can also be obtained by using the distributions of fig. 5.5(*b*) and 5.5(*d*). (Equations (5.8) and (5.9) will have to be used to get the answer into the same form.)

5.2 A uniform beam, full plastic moment M_p and flexural rigidity EI, is pinned to four rigid supports as shown. The point load W acts at the mid-point of the central span, and is gradually increased until collapse occurs at the value $8M_p/l$. Find the value of W at which the first plastic hinge forms, and

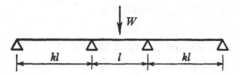

estimate the corresponding deflexion at the loading point. Determine also the deflexion of the load at incipient collapse. (Ignore both strain hardening and the spread of the plastic zones.)

$$\left(Ans. \ \frac{8M_p}{l}\left(\frac{2k+3}{4k+3}\right), \ \frac{M_p l^2}{24EI}\left(\frac{8k+3}{4k+3}\right), \ \frac{M_p l^2}{24EI}(4k+1).\right)$$

5.3 (Stüssi paradox.) Sketch the load deflexion curves for the problem of example 5.2, using the given answers, for $k = 0, 1, 2, 4, \infty$.

5.4 Check the values of the loads W given in equations (9.1) as the frame of fig. 9.1 is loaded slowly to collapse. Determine also the sidesway deflexion of the frame at the point of formation of each of the four hinges.

$$\left(Ans. \ \delta \left[= \frac{l^2}{6EI}(2M_A + M_B)\right] = \frac{35}{33}\left(\frac{M_p l^2}{6EI}\right), \ \frac{79}{67}\left(\frac{M_p l^2}{6EI}\right),\right.$$
$$\left.\frac{41}{23}\left(\frac{M_p l^2}{6EI}\right), \ 2\left(\frac{M_p l^2}{6EI}\right).\right)$$

5.5 The frame shown has uniform section of full plastic moment M_p and flexural rigidity EI (cf. fig. 4.12 in vol. 1). Confirm that the value of W at collapse is given by $W_c = M_p/l$. Determine the sidesway deflexion at the level of the upper beam at incipient collapse, and the corresponding value of the bending moment at the cross-section marked X.

$$\left(Ans. \ \frac{25}{24}\frac{M_p l^2}{EI}, \ -\tfrac{3}{8}M_p.\right)$$

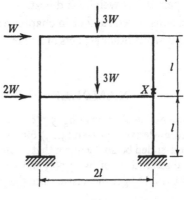

5.6 The frame of example 5.5 collapses by the formation of only six hinges. Using the method of combination of mechanisms for an artificially strengthened frame, determine a statically admissible value for the bending moment at the cross-section X. ($Ans.\ -\frac{1}{2}M_p.$)

5.7 Confirm that the frame shown collapses at a load factor 1·976 (cf. figs. 4.20 and 4.24 (p) of vol. 1). By considering series of artificially strengthened frames, derive the statically admissible bending-moment distribution whose numerical values at collapse are shown adjacent to the indicated hinge points.

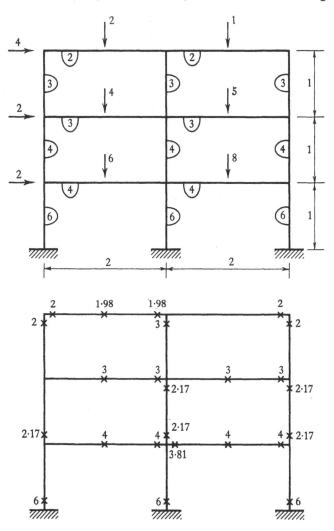

6

REPEATED LOADING

In all the work in both vols. 1 and 2 it has been assumed that the *relative* magnitudes of the loads acting on a frame have been fixed. It is true that there has been, if only implicitly, some possibility of choice of combination of loads; for example, the wind blowing from left to right might give a more critical loading case than loading in the opposite direction. Similarly, it has been assumed that the designer will have arranged other live loads in the worst possible way, that is, in the way which will lead to the largest values of full plastic moment of the members. For a complex structure more than one pattern of loading may have to be considered; for example, one design might be carried out under dead plus superimposed loading at a certain load factor, say 1·75, and another under dead plus superimposed plus wind loading at a lower factor, say 1.4. The designer would take the more critical of these two cases for any particular structure.

However, once a critical pattern (or patterns) of loading has been fixed, the problem has then been treated as one of *statical* collapse. For the purposes of analysis, all loads are imagined to be increased slowly and in proportion until collapse of the frame occurs by the formation of a mechanism of collapse at a load factor λ_c. For design, the required load factor is first applied to all the loads, and the frame is then proportioned just to collapse under these factored loads.

This situation is not altered if a more sophisticated approach is adopted in which different classes of loads (say dead and live loads) are assigned different load factors. The essential feature of this kind of design process is that, once the pattern of loading has been selected, the individual loads contributing to that pattern are not supposed to be independent of each other but, on the contrary, are assumed to be given in terms of a single parameter. The present chapter is concerned with loading patterns in which the various loads can act randomly and independently within given limits. These limits might simply correspond to the maximum value of a particular load and zero; there may be snow on the roof of a building, whose weight is specified in the appropriate Standard, or there may be no snow. In the case of wind loading, the lower limit might well be negative, corresponding to the wind blowing in one direction rather than the other.

Thus, instead of working *values* of the loads, there will in practice be a working *range* of values. Moreover, the current value of one particular load within its own range will, in general, be independent of the values of all the other loads. There are two main effects to be considered of the consequent random and repeated loading of a structure. The first, that of alternating plasticity, is easy to analyse. It may be that under a certain combination of the independently varying loads a plastic hinge develops, either partially or fully, at a certain cross-section of the frame. At a later time, a different combination of loads may produce plasticity at the same cross-section but with the bending moment acting in the opposite sense. Such repeated bending back and forth at a cross-section may not be very harmful if the number of repetitions is fairly small, but it is easy, as will be seen, to compute the permitted range of loading to prevent its occurrence. The phenomenon is, in any case, usually less critical than that of incremental collapse, which is the second main effect of repeated loading.

It was noted above that a certain combination of the independently varying loads might cause a plastic hinge to form at a certain cross-section. If so, then some small amount of rotation will take place at that hinge position; the rotation will be small because a single hinge (or two or three hinges) will not produce a collapse mechanism, so that any plastic deformation must be *contained* by those portions of the frame remaining elastic. Under a different combination of loads it is possible that another hinge or hinges could form at different cross-sections, and again some small amounts of plastic rotation could occur. A third combination of loads could lead to the formation of yet other plastic hinges, and so on.

Now although collapse may not occur under any *single* combination of loads (because the number of hinges involved is too small to make a complete mechanism), it may be that all the hinges formed under *different* loading combinations would correspond to a mechanism had they all occurred simultaneously. If this is so, then a mode of incremental collapse is possible. As a simple qualitative example (which, however, can be confirmed for certain numerical values of loads and dimensions), the simple rectangular frame will be considered. The usual idealized loads V and H will be taken as varying independently between zero and specified maximum values V_{max} and H_{max}.

In fig. 6.1 (a), the side load is shown acting alone, and it is possible for hinges to be formed at sections A and B of the frame. The consequent small rotations at these hinges are not recoverable; on unloading the frame, the response is elastic, and column AB will no longer be vertical, but will have a slight inclination. If the frame is now reloaded with both vertical and side loads, fig. 6.1 (b), hinges can form at sections D and E, again with plastic rotations occurring at these points; after unloading, column DE will be left with a slight inclination to the vertical.

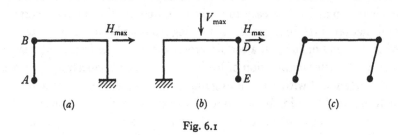

Fig. 6.1

The cycle may now be repeated. Upon reloading with the side load only, hinges A and B will form again, and column AB will become further inclined. After a very few repetitions of such cyclic loading (say ten repetitions or less) the frame will have the appearance sketched in fig. 6.1 (c); it will look as if it is failing by a well-established sidesway mode. This type of incremental collapse will occur if the magnitudes of V_{max} and H_{max} exceed certain calculable values. For smaller values of the loads, then, it is possible that some plastic deformation may occur in the first few cycles of loading, but it is also possible that thereafter all further changes of load are resisted purely elastically by the frame. If this happens, then the frame is said to have shaken down under the variable repeated loading.

There is evidently some limit (the shakedown limit) which divides shakedown behaviour from that of incremental collapse (or alternating plasticity), and it is convenient to retain the idea of a load factor in order to assign a numerical value to this limit. The load factor is applied not to the current values of the loads, but to the ranges within which the loads can act. Thus, in fig. 6.1 (b), the loads V_{max} and H_{max} might represent the working values (taken from the appropriate Standard) of snow and wind loading respectively; more generally, the actual values of the loads will be assumed to be contained at all times within the ranges

$$\left.\begin{array}{c} V_{min} \leqslant V \leqslant V_{max}, \\ H_{min} \leqslant H \leqslant H_{max}. \end{array}\right\} \tag{6.1}$$

A satisfactory design of frame must necessarily be one that resists purely elastically all variations of the loads within the ranges (6.1); even if some initial plastic deformation occurs, the frame eventually shakes down. (There may, of course, be more severe design criteria for any particular frame.) If the ranges of loading are widened by a factor λ, however, so that

$$\left. \begin{array}{l} \lambda V_{\min} \leqslant V \leqslant \lambda V_{\max}, \\ \lambda H_{\min} \leqslant H \leqslant \lambda H_{\max}, \end{array} \right\} \tag{6.2}$$

and the value of λ is imagined to be slowly increased, then the frame will eventually no longer be able to shake down, and either incremental collapse or alternating plasticity will occur. The limiting value of λ, denoted λ_s, is the shakedown load factor.

It is clear that for small enough values of λ there will be no plastic deformation as the loads vary independently within their ranges; an initially stress-free frame will remain stress free at any instant when all the loads happen to be removed. As the value of λ is increased, a stage will be reached at which a single critical section (or, by numerical accident, two or more critical sections) will just begin to yield under the most unfavourable combination of the independent loads. The corresponding value of λ is, effectively, the conventional safety factor on stress. Just as for the static case, the mere attainment of yield at one or more cross-sections does not imply collapse; the value of λ could be increased and the frame might still be able to shake down. However, an increase in the value of λ above that of the conventional safety factor does, by definition, imply that yield is occurring somewhere in the frame. Such yield in turn implies that, if all the loads were removed, the frame would no longer be stress free; the frame has been distorted locally and self-stressing (residual) moments must have been induced. It is, of course, the presence of these residual moments that allows the frame to shake down for values of λ below λ_s, but above the value of the conventional safety factor.

As a very simple example, the fixed-ended beam of section 5.2 may be investigated under the action of a load W in the range

$$0 \leqslant W \leqslant W_{\max}. \tag{6.3}$$

From equations (5.10), no yield whatever will occur if $W_{\max} \leqslant 9M_p/4l$ (it is assumed that the beam is perfectly encastré, and that the moment–curvature relationship is elastic–perfectly plastic, fig. 5.1). However, if (for example) the value of $W_{\max}$ were set at $2 \cdot 7 M_p/l$, then the (hypothetical) elastic moments in the beam under the action of the full load would be as shown in fig. 6.2(b); the moment at end C of the beam is shown having the apparently impossible value of $1 \cdot 2 M_p$. On the first application of the full load, therefore, a hinge would form at end C, and the bending-moment distribution would be that of fig. 6.2(c); the values marked are taken from equations (5.11) for $W = 2 \cdot 7 M_p/l$.

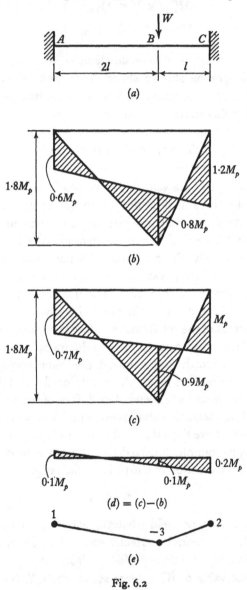

Fig. 6.2

130

Upon the *removal* of the load W there is nothing to prevent the response from being completely elastic. As the value of W is decreased from $2.7M_p/l$ to zero, the bending moment at end C of the beam will decrease from the value M_p to $-0.2M_p$. Complete removal of the external loading will leave a set of self-stressing moments in the beam, whose magnitudes, fig. 6.2(d), are given simply by the differences of the bending-moment diagrams of fig. 6.2(b) and (c). These residual moments, induced by plastic deformation on first loading, will enable further loading to be resisted elastically; the favourable negative moment of $-0.2M_p$ will allow the superposition of an elastic response at end C of magnitude $1.2M_p$ before yielding again starts. The beam has shaken down after a single load application; in fact in this simple example, the collapse load and the limiting load for shakedown are the same.

It should be noted that the residual moments m of fig. 6.2(d),

$$\{m_A, m_B, m_C\} \equiv \{0.1M_p, -0.1M_p, -0.2M_p\}, \tag{6.4}$$

plot to give a linear bending-moment diagram, as they should if they are to form a proper set of self-stressing moments. Since the moments m are in equilibrium with zero external load, then *any* mechanism θ will give the virtual work relationship

$$\Sigma m\theta = 0. \tag{6.5}$$

Thus the virtual mechanism of fig. 6.2(e) taken with set (6.4) will satisfy equation (6.5).

Further, the residual pattern of fig. 6.2(d), once established, remains unchanged on further loading. It will be seen later that if a structure *is* to shake down, then any residual moments that may be induced by the loading must eventually become stabilized in this way. However, more complex loading systems, involving the independent variation of two or more loads, may cause varying patterns of residual moments for the first few loading cycles. It is the presence of residual moments that destroys the symmetry of the response of the portal frame in fig. 6.1. The formation of hinges A and B, fig. 6.1(a), will induce certain residual moments in the frame. Under the loading of fig. 6.1(b), in which hinges form at D and E, the pattern of residual moments is altered, leaving a different state of self stress when the loads are removed. This new state makes sections A and B more critical than D and E if the side load alone is again applied; if the loads have magnitudes above the shakedown limit, there will be a continual shifting in the values of the residual moments.

6.1 The shakedown theorem

The discussion above assumed, for simplicity, a material that was either elastic or perfectly plastic; the moment–curvature relationship consisted of straight lines. The shakedown theorem can be established for a material having the more general moment–curvature relationship of fig. 6.3. The basic curve (OAB) is assumed to be the same whichever way the bending moment is applied, so that first yield occurs at a moment M_y in either sense, and the full plastic moment has value M_p, again in either sense. The linear elastic range thus extends for a total $2M_y$. The assumption is made that this range of $2M_y$ is not affected by any partially plastic deformation that might occur. Thus if a moment corresponding to the value at point B in fig. 6.3 is applied to the cross-section, followed by unloading, then the behaviour will be linear for a total decrease of moment $2M_y$.

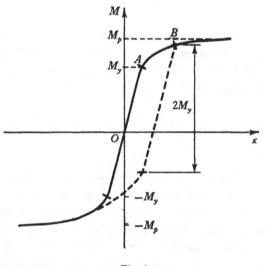

Fig. 6.3

The necessary conditions for shakedown can now be written in terms of the conventional elastic solution for a frame and of the residual moments which may exist. It will be appreciated that both these components must enter the formulation of the problem. The elastic solution is required, since the value of λ_s is sought for which the response of the frame is entirely elastic. On the other hand, since some plastic deformation can occur, residual moments will be induced which will affect the total value of the bending moment at any cross-section.

Using the working values of the loads, the value $\mathcal{M}_i$ of the elastic bending moment may be computed for each critical section i of the frame. As the individual loads vary between their prescribed maximum and minimum values, so the bending moment $\mathcal{M}_i$ will vary, and the greatest and least values $\mathcal{M}_i^{\max}$ and $\mathcal{M}_i^{\min}$ may be evaluated. A load factor λ applied to the *range* of loading will increase these values to $\lambda\mathcal{M}_i^{\max}$ and $\lambda\mathcal{M}_i^{\min}$. These factored values of the elastic bending moments are those that would occur if the frame remained undistorted, but there will in general be a residual moment m_i at the cross-section which must be added to the elastic values to give the total bending moment at that section. Thus the necessary conditions for shakedown to occur are that

$$\left.\begin{aligned}\lambda\mathcal{M}_i^{\max}+m_i &\leqslant (M_p)_i, \\ \lambda\mathcal{M}_i^{\min}+m_i &\geqslant -(M_p)_i,\end{aligned}\right\} \tag{6.6}$$

and, to guard against the possible danger of alternating plasticity,

$$\lambda(\mathcal{M}_i^{\max}-\mathcal{M}_i^{\min}) \leqslant 2M_y. \tag{6.7}$$

The inequality (6.7) is clearly sufficient as well as necessary for the avoidance of alternating plasticity. The shakedown theorem states that if *any* set of residual moments $\bar{m}_i$ can be found to satisfy inequalities (6.6) at a load factor λ, then shakedown will occur at that load factor; that is, inequalities (6.6) are sufficient as well as necessary. These two inequalities control the phenomenon of incremental collapse, as will be seen from some simple numerical examples below.

(Inequality (6.7) is not usually of critical importance in any practical design of a framed structure, although it may be noted here that this is not necessarily true for problems involving more general stress systems. The plastic design of pressure vessels, for example, may be governed by alternating plasticity. This is because stress concentration factors of 3 or more are not all uncommon, so that, with a collapse load factor of, say, 2, first yield may well occur long before the full *working* values of the loads are applied. For the usual types of framed structure, however, designed to a load factor of, say, 1.5 or 2, it is not likely that yield will occur at a value of the load factor below unity, unless this is accidental, due to settlement of supports or other imperfections. Mere removal of the load from such a frame would not therefore cause yield in the reverse direction, although this could well happen with a pressure vessel.)

To establish the shakedown theorem, a small change will be considered in the values of the applied loads. If the residual moment at any section of the frame has value m_i at a given instant, then the variation of the applied loads may cause some yielding to occur, changing the value of the residual moment to $m_i + \delta m_i$. During this process, the change in curvature $\delta m_i / EI$ at each portion of the frame which remains elastic will be compatible with any hinge rotations $\delta \theta_k$ which may occur at the yielding sections k. Thus since m_i itself is a set of bending moments in equilibrium with zero external load, the equation of virtual work gives

$$\int m_i \frac{\delta m_i}{EI} \, ds + \Sigma m_k \delta \theta_k = 0, \qquad (6.8)$$

where the integration extends over all portions of the frame which remain elastic during the small changes of applied loading, and the summation includes all hinge rotations which occur.

A set of residual moments satisfying inequalities (6.6) will be denoted $\overline{m}_i$. A second application of the equation of virtual work then gives

$$\int \overline{m}_i \frac{\delta m_i}{EI} \, ds + \Sigma \overline{m}_k \delta \theta_k = 0, \qquad (6.9)$$

and, combining equations (6.8) and (6.9),

$$\int (m_i - \overline{m}_i) \frac{\delta m_i}{EI} \, ds + \Sigma (m_k - \overline{m}_k) \delta \theta_k = 0. \qquad (6.10)$$

Suppose that at a particular section k where yield is occurring, the current value of m_k is such that $m_k < \overline{m}_k$, that is

$$(m_k - \overline{m}_k) < 0. \qquad (6.11)$$

The first of inequalities (6.6) then gives

$$\lambda \mathcal{M}_k^{\max} + \overline{m}_k \leqslant (M_p)_k, \Big\}$$

that is

$$\lambda \mathcal{M}_k^{\max} + m_k < (M_p)_k, \Big\} \qquad (6.12)$$

where, in the second inequality, the possibility of equality no longer exists. Thus, since yield *is* occurring *ex hypothesi* at the section, it must be under the negative value of full plastic moment $-(M_p)_k$, and the corresponding value $\delta \theta_k$ of the hinge rotation must also be negative. Thus, using (6.11),

$$(m_k - \overline{m}_k) \delta \theta_k > 0. \qquad (6.13)$$

Similarly, if the assumption is made that $m_k > \overline{m}_k$, then it may be concluded that $\delta \theta_k > 0$. If $m_k = \overline{m}_k$, then no information results as to the sign of $\delta \theta_k$, but in all cases

$$(m_k - \overline{m}_k) \delta \theta_k \geqslant 0. \qquad (6.14)$$

Hence, from equation (6.10),

$$\int (m_i - \overline{m}_i)\frac{\delta m_i}{EI}\,ds \leqslant 0. \tag{6.15}$$

Now the quantity

$$U = \int \frac{(m_i - \overline{m}_i)^2}{2EI}\,ds \tag{6.16}$$

is positive definite, and inequality (6.15) states that, as the loading on the frame changes, $\delta U \leqslant 0$. Thus the value of U decreases if any plastic deformation is occurring, and remains constant otherwise; the value of U must either become zero (in which case $m_i = \overline{m}_i$ everywhere) or must settle down to a definite positive value. In either case the values of m_i become constant, so that shakedown occurs.

From this outline proof of the shakedown theorem, it will be evident that (just as for static collapse) the presence of an *initial* set of residual stresses will have no effect on the final behaviour of the frame. A frame which is initially imperfect, or which suffers subsequent settlement, will be subject to a set of residual moments which may well cause hinges to form in an order different from that predicted by the simple theory. However, an initial pattern of residual moments will be 'wiped out' by later yielding, and the limiting load factor λ_s for shakedown will have a unique value.

6.2   A two-span beam

As a first numerical example, the two-span beam of fig. 6.4(a) will be investigated. The continuous beam is of uniform section, full plastic moment M_p and yield moment M_y, and the loads W_1 and W_2 vary independently between the identical limits

$$\left.\begin{array}{l} 0 \leqslant W_1 \leqslant W_0, \\ 0 \leqslant W_2 \leqslant W_0. \end{array}\right\} \tag{6.17}$$

The largest value of W_0 is sought for which the beam system will just shake down. (For later comparison, the *static* value of W_0 is given by $W_0 = W_c = 6M_p/l$.)

The basic elastic solution for a single load ($W_1 = 0, W_2 = W_0$) is shown in fig. 6.4(b); the value of the redundant moment x is given by any usual method of elastic analysis as $x = \frac{3}{32}W_0 l$, so that the corresponding value of the bending moment under the load is $-\frac{13}{64}W_0 l$. Table 6.1 may now be drawn up.

Table 6.1

Section	Due to W_1		Due to W_2		Combined loading		
	$\mathcal{M}^{\max}$	$\mathcal{M}^{\min}$	$\mathcal{M}^{\max}$	$\mathcal{M}^{\min}$	$\mathcal{M}^{\max}$	$\mathcal{M}^{\min}$	$\mathcal{M}^{\max}-\mathcal{M}^{\min}$
A	0	$-\frac{11}{64}W_0 l$	$\frac{3}{64}W_0 l$	0	$\frac{3}{64}W_0 l$	$-\frac{11}{64}W_0 l$	$\frac{14}{64}W_0 l$
B	$\frac{6}{64}W_0 l$	0	$\frac{6}{64}W_0 l$	0	$\frac{12}{64}W_0 l$	0	$\frac{12}{64}W_0 l$
C	$\frac{3}{64}W_0 l$	0	0	$-\frac{11}{64}W_0 l$	$\frac{3}{64}W_0 l$	$-\frac{11}{64}W_0 l$	$\frac{14}{64}W_0 l$

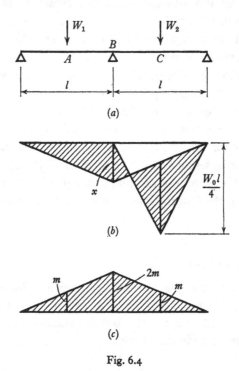

Fig. 6.4

From the values of maximum and minimum bending moment under combined loading, it is clear that shakedown will certainly occur for

$$\frac{13}{64}W_0 l \leqslant M_p,\qquad(6.18)$$

that is, $W_0 \leqslant 4\cdot92 M_p/l$, cf. $W_c = 6M_p/l$. (The question of alternating plasticity will be deferred for the moment.) However, the writing of inequalities (6.6) for sections A and B (the behaviour of section C is the same as that of A) will lead to a higher limit for W_0. The most general pattern of residual moments is sketched in fig. 6.4(c), with the value of the residual moment at B just twice that at A. Thus the inequalities became

$$\left.\begin{array}{r}\tfrac{3}{64}W_0 l + m \leqslant M_p, \\[4pt] -\tfrac{13}{64}W_0 l + m \geqslant -M_p, \\[4pt] \tfrac{12}{64}W_0 l + 2m \leqslant M_p, \\[4pt] 0 + 2m \geqslant -M_p.\end{array}\right\} \tag{6.19}$$

There are systematic ways of dealing with sets of linear inequalities such as (6.19), one of which is discussed briefly in chapter 8. However, this particularly simple set can be examined by trial and error. Since there are two unknown quantities (m and M_p), the equality sign may be taken in *two* of (6.19). For example, if the first two of (6.19) are taken as equalities, they give the values $m = \tfrac{5}{64}W_0 l$, $M_p = \tfrac{8}{64}W_0 l$; however, the third of (6.19) then becomes $\tfrac{22}{64}W_0 l \leqslant \tfrac{8}{64}W_0 l$. It turns out that the second and third of the inequalities are critical, corresponding to a sagging hinge under the loading points A and C and a hogging hinge at the support B. The values

$$m = \tfrac{1}{192}W_0 l, \quad M_p = \tfrac{19}{96}W_0 l \tag{6.20}$$

just lead to equalities in the second and third of (6.19), while the first and last are easily satisfied. Thus the shakedown limit under these conditions is $W_0 = 5\cdot05 M_p/l$, which is a reduction of almost 16 per cent from the static value of the collapse load.

With the value of W_0 given by (6.20), the largest range of bending moment, $\mathcal{M}^{\max} - \mathcal{M}^{\min}$, is, from table 6.1, $\tfrac{24}{19}M_p = 1\cdot26 M_p$. Thus, providing M_y exceeds $0\cdot63 M_p$ (as it certainly will for a rolled I-section whose shape factor is about $1\cdot15$), there is no danger of alternating plasticity.

The same example may be altered slightly to illustrate the behaviour of a frame which is collapsing incrementally. The value of W_1 will be fixed at $5\cdot6 M_p/l$, and W_2 will be allowed to fluctuate between zero and the same value $5\cdot6 M_p/l$. For simplicity, the arguments will be based on the moment–curvature relationship consisting of two straight lines, but the resulting conclusions would be very little altered if a full analysis were made for the more realistic relationship of fig. 6.3. On first loading with W_1 up to the full fixed value $5\cdot6 M_p/l$, yield will occur (under the load) at $4\cdot92 M_p/l$ (cf. (6.18)). As W_1 is increased above this value, some rotation will occur at this hinge, and the total discontinuity may be calculated (if required) by the methods of chapter 5. At full load the bending-moment distribution, which is statically determinate due to the presence of the hinge, will be as shown in fig. 6.5 (*a*).

The value of W_2 is now slowly increased from zero, while W_1 remains fixed. The bending moment under the load W_1 is reduced in value, and, when W_2 has reached such a value that

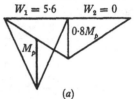

$$\tfrac{3}{32}W_2 l = 0.2 M_p$$

(that is, $W_2 = 2.13 M_p/l$), a plastic hinge just forms at the central support, fig. 6.5 (b). At this stage, the bending moment under the load W_1 is $0.9 M_p$. As W_2 is increased from $2.13 M_p/l$ to its full value $5.6 M_p/l$, the central hinge rotates; during this process, the bending-moment diagram for the left-hand span remains fixed, fig. 6.5 (c).

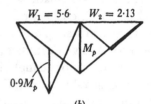

The value of W_2 is now decreased slowly from its full value, while W_1 still remains fixed. The initial response of the whole beam is again entirely elastic, since the moment at the single active hinge is reduced. When W_2 has been reduced by $2.13 M_p/l$ to $3.47 M_p/l$, the moment under the left-hand load W_1 has increased to the value M_p, fig. 6.5 (d), and further reduction of W_2 is accompanied by plastic hinge rotation at this section. When W_2 has been completely removed, the bending moments in the beam will be as in fig. 6.5 (e), which is identical with fig. 6.5 (a), and the cycle is ready to be repeated.

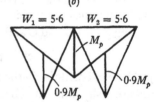

Figure 6.5 gives an example of incremental collapse under fixed loading due to varying loads elsewhere on the structure. Starting from fig. 6.5 (a), each time the load W_2 is applied the central hinge rotates, and each time W_2 is removed the hinge under

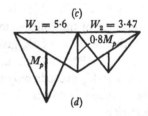

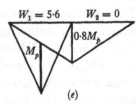

Fig. 6.5

W_1 rotates. After a very few cycles the *left*-hand span would show a well-developed collapse mechanism due to variation of loading in the *right*-hand span.

6.3 The combination of mechanisms

The writing of inequalities such as (6.19) in order to determine a shake-down solution is analogous to the combination of free and reactant bending moments to determine static collapse loads of frames. The two-span beam has a single redundancy, so that one residual moment was introduced; a similar solution for a fixed-base portal frame would intro-duce three separate residual moments, which could then be eliminated from the inequalities to leave a single shakedown relation.

To show even more clearly the exact correspondence of the shakedown and static collapse solutions, the *elastic* moments in table 6.1 will be used to determine the *static plastic* collapse load for the two-span beam, assuming W_1 and W_2 to both have their full values W_c. Since both W_1 and W_2 act together, the static elastic moments are, from table 6.1,

$$\left.\begin{aligned}
\mathcal{M}_A &= -\tfrac{10}{64}W_c l, \\
\mathcal{M}_B &= \tfrac{12}{64}W_c l, \\
\mathcal{M}_C &= -\tfrac{10}{64}W_c l.
\end{aligned}\right\} \tag{6.21}$$

On superimposing the residual distribution of fig. 6.4(c), the inequalities for sections A and B, analogous to those of (6.19), become

$$\left.\begin{aligned}
M_p &\geqslant -\tfrac{10}{64}W_c l + m \geqslant -M_p, \\
M_p &\geqslant \tfrac{12}{64}W_c l + 2m \geqslant -M_p.
\end{aligned}\right\} \tag{6.22}$$

These continued inequalities can again be dealt with by trial and error; for a given value of M_p, the largest value of W_c will result when

$$m = -\tfrac{1}{96}W_c l, \quad M_p = \tfrac{1}{8}W_c l. \tag{6.23}$$

Thus W_1 and W_2 may be increased slowly and simultaneously to a collapse value of $6M_p/l$; upon unloading, a residual (sagging) moment will be found at the central support of magnitude $2m = -\tfrac{1}{8}M_p$.

The formulation of (6.21) and (6.22) shows at once that the procedure is wasteful if only the collapse equation is required. The value of the residual moment m was given in (6.23), but this value is usually not required. The mechanism approach to static collapse analysis deals directly with free bending moments and with the values of the full plastic moments, and makes no reference to the values of the redundancies. Similarly, the mechanism equations for incremental collapse analysis do not involve the values of the residual moments. The price paid for this simplification of the working is exactly the same in the static and in the incremental collapse cases; the correct solution is not derived automatically. Instead, different mechanisms have to be investigated to find which one gives the smallest value of the load factor; that is, the unsafe theorem carries over from static collapse to incremental collapse.

Suppose that the correct incremental collapse mechanism is known for a given frame so that hinge rotations ϕ_i can be assigned at a certain number of critical sections. If the value of ϕ_i at any particular section is positive, then that hinge will form when the total bending moment is equal to $+M_p$. The first of inequalities (6.6) will then be replaced by the equality

$$\lambda_s \mathcal{M}_i^{\max} + m_i = (M_p)_i. \tag{6.24}$$

The corresponding rotation of the incremental collapse mechanism will be denoted ϕ_i^+. Similarly, if the value of the hinge rotation at the section were negative, denoted $-\phi_i^-$, where ϕ_i^- is itself a positive quantity, then

$$\lambda_s \mathcal{M}_i^{\min} + m_i = -(M_p)_i. \tag{6.25}$$

If now equations (6.24) and (6.25) are multiplied by the hinge rotation ϕ_i, then either

$$\left. \begin{aligned} \lambda_s \mathcal{M}_i^{\max} \phi_i^+ + m_i \phi_i &= (M_p)_i |\phi_i|, \\ -\lambda_s \mathcal{M}_i^{\min} \phi_i^- + m_i \phi_i &= (M_p)_i |\phi_i|, \end{aligned} \right\} \tag{6.26}$$

or

where the first equation is taken if the value of ϕ_i is positive, and the second if ϕ_i is negative. The product $M_p \phi$ on the right-hand side is, as usual, always positive.

Summing equations (6.26) for all hinges of the collapse mechanism,

$$\lambda_s [\Sigma \mathcal{M}_i^{\max} \phi_i^+ - \Sigma \mathcal{M}_i^{\min} \phi_i^-] + \Sigma m_i \phi_i = \Sigma (M_p)_i |\phi_i|. \tag{6.27}$$

Now, since the m_i are residual moments in equilibrium with zero external load, and the ϕ_i are hinge rotations of a proper mechanism, then $\Sigma m_i \phi_i$ is zero by virtue of the virtual work equation. The basic equation for incremental collapse then becomes

$$\lambda_s [\Sigma \mathcal{M}_i^{\max} \phi_i^+ - \Sigma \mathcal{M}_i^{\min} \phi_i^-] = \Sigma (M_p)_i |\phi_i|. \tag{6.28}$$

As an example of the use of this equation, the two-span beam, fig. 6.4 and table 6.1, collapses in the incremental mechanism of fig. 6.6 (with, of course, simultaneous collapse in the other span, since the problem is symmetrical). Since the rotation at A is negative, the value $\mathcal{M}_A^{\min}$ must be taken from table 6.1; similarly, the value $\mathcal{M}_B^{\max}$ must be used. Writing equation (6.28) in full, and treating the problem as one of design for unit load factor λ_s,

$$[-2\mathcal{M}_A^{\min} + \mathcal{M}_B^{\max}] = 3M_p,$$

or

$$[\tfrac{13}{32}W_0 l + \tfrac{6}{32}W_0 l] = 3M_p; \qquad (6.29)$$

equation (6.29) leads, of course, to the value of M_p given previously in (6.20). If the value of the residual moment m is required for any reason (as it may be, in general, to make a statical check of the frame), then it may be found by writing the original equations (6.24) or (6.25). Thus the second of (6.19), for example, gives

$$-\tfrac{13}{64}W_0 l + m = -M_p = -\tfrac{19}{96}W_0 l,$$

or

$$m = \tfrac{1}{192}W_0 l. \qquad (6.30)$$

A more complex example of this sort of analysis is given below for the rectangular portal frame.

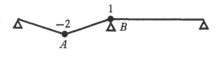

Fig. 6.6

The incremental mechanism equation, (6.28), will usually be written for an *assumed* mechanism of collapse; that is, the correct collapse mode for a complex frame will not be known by inspection, but has to be determined. The method of combination of mechanisms can be used for this purpose. The value of the load factor for incremental collapse which results from an assumed mechanism is always an upper bound to the correct value; the proof follows the same lines as that for the corresponding static theorem (vol. 1, p. 134). When combining mechanisms all the previous techniques can be used, but there is one small and significant change; work terms must be subtracted from *both* sides of mechanism equations, similar to (6.28), to allow for the cancellation of a hinge.

This may be seen by considering two independent mechanisms ϕ and ψ, which are superimposed in order to cancel a particular hinge, say at section A of the frame. If the hinge is to disappear in the combined mechanism, then the rotation $\phi_A^+ = \theta_A$ must just cancel a rotation $-\psi_A^- = -\theta_A$, where (arbitrarily) the hinge rotation at A has been supposed positive for mechanism ϕ.] Writing the two mechanism equations with the contributions from hinge A displayed separately,

$$\lambda[\mathscr{M}_A^{\max}\theta_A + \sum_{\text{Not }A} (\mathscr{M}_i^{\max}\phi_i^+ - \mathscr{M}_i^{\min}\phi_i^-)] = (M_p)_A\theta_A + \sum_{\text{Not }A} (M_p)_i|\phi_i|,$$

$$\lambda[-\mathscr{M}_A^{\min}\theta_A + \sum_{\text{Not }A} (\mathscr{M}_i^{\max}\psi_i^+ - \mathscr{M}_i^{\min}\psi_i^-)]$$
$$= (M_p)_A\theta_A + \sum_{\text{Not }A} (M_p)_i|\psi_i|.$$

$$(6.31)$$

These two equations may be added to represent the combined mechanism with no hinge at A:

$$\lambda[(\mathscr{M}_A^{\max} - \mathscr{M}_A^{\min})\theta_A + \sum_{\text{Not }A} \{\mathscr{M}_i^{\max}(\phi_i^+ + \psi_i^+) - \mathscr{M}_i^{\min}(\phi_i^- + \psi_i^-)\}]$$
$$= 2(M_p)_A\theta_A + \sum_{\text{Not }A} (M_p)_i|\phi_i + \psi_i|. \quad (6.32)$$

Now there should be no contributions in this equation from the hinge at A, so that, as usual, $2(M_p)_A\theta_A$ should be subtracted from the right-hand side. In addition, however, the term $(\mathscr{M}_A^{\max} - \mathscr{M}_A^{\min})\theta_A$ should be subtracted from the left-hand side; a work term due to the *range* of elastic moment at A disappears when the mechanisms are combined.

It should be noted that, exactly as for the static case, it has been assumed implicitly that the signs of the hinge rotations in the two mechanisms are the same at any particular section, apart from the section A where the hinge is being cancelled. In the last term of equation (6.32) for example, $|\phi_i + \psi_i|$ is only equal to $|\phi_i| + |\psi_i|$ if the signs of ϕ_i and ψ_i are either both positive or both negative; a similar remark applies to the corresponding terms on the left-hand side of the equation.

(It may be noted that a static collapse analysis may be made directly if an elastic solution is available. For proportional loading there is a single value $\mathscr{M}_i$ of the elastic bending moment at each critical section. If a hinge forms at section i, then, corresponding to equations (6.24) and (6.25), the single equation

$$\lambda_c\mathscr{M}_i + m_i = \pm(M_p)_i \qquad (6.33)$$

may be written. The arguments then follow as before, and equation (6.28) is replaced by

$$\lambda_c\Sigma\mathscr{M}_i\phi_i = \Sigma(M_p)_i|\phi_i|. \qquad (6.34)$$

This method is, of course, laborious unless the elastic solution $\mathscr{M}_i$ is already available; otherwise *any* set of equilibrium moments $(M_F)_i$ can be used on the left-hand side of equation (6.34) in order to calculate λ_c; (cf. equation (3.11) of vol. 1.)

6.4 Rectangular portal frames

The pinned-base rectangular frame of fig. 6.7 has uniform section, full plastic moment M_p. The three possible modes of collapse (for positive values of V and H) are shown. The usual static collapse equations are

$$
\begin{aligned}
(a) &\quad Hh &= 2M_p, \\
(b) &\quad Hh + \tfrac{1}{2}Vl = 4M_p, \\
(c) &\quad \tfrac{1}{2}Vl = 4M_p.
\end{aligned}
\tag{6.35}
$$

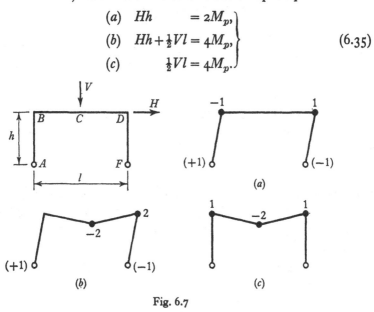

Fig. 6.7

Equations similar to (6.35) are sought for the case when the loads are no longer static, but can vary independently within the limits

$$
\begin{aligned}
0 &\leqslant V \leqslant V_0, \\
0 &\leqslant H \leqslant H_0.
\end{aligned}
\tag{6.36}
$$

The basic elastic solution for the frame is given in table 6.2.

Table 6.2

Section	Due to V		Due to H		Combined loading	
	$\mathcal{M}^{\max}$	$\mathcal{M}^{\min}$	$\mathcal{M}^{\max}$	$\mathcal{M}^{\min}$	$\mathcal{M}^{\max}$	$\mathcal{M}^{\min}$
B	$\dfrac{V_0 l}{8}\left(\dfrac{3l}{2h+3l}\right)$	0	0	$-\dfrac{H_0 h}{2}$	$\dfrac{V_0 l}{8}\left(\dfrac{3l}{2h+3l}\right)$	$-\dfrac{H_0 h}{2}$
C	0	$-\dfrac{V_0 l}{8}\left(\dfrac{4h+3l}{2h+3l}\right)$	0	0	0	$-\dfrac{V_0 l}{8}\left(\dfrac{4h+3l}{2h+3l}\right)$
D	$\dfrac{V_0 l}{8}\left(\dfrac{3l}{2h+3l}\right)$	0	$\dfrac{H_0 h}{2}$	0	$\dfrac{V_0 l}{8}\left(\dfrac{3l}{2h+3l}\right)+\dfrac{H_0 h}{2}$	0

Thus using equation (6.28), with $\lambda_s = 1$, with the three mechanisms of fig. 6.7,

(a) $\quad \left[\dfrac{H_0 h}{2}\right] + \left[\dfrac{V_0 l}{8}\left(\dfrac{3l}{2h+3l}\right) + \dfrac{H_0 h}{2}\right] = 2M_p,$

or $\quad\quad\quad\quad\quad H_0 h + \dfrac{V_0 l}{8}\left(\dfrac{3l}{2h+3l}\right) = 2M_p;$

(b) $\quad 2\left[\dfrac{V_0 l}{8}\left(\dfrac{4h+3l}{2h+3l}\right)\right] + 2\left[\dfrac{V_0 l}{8}\left(\dfrac{3l}{2h+3l}\right) + \dfrac{H_0 h}{2}\right] = 4M_p,$

or $\quad\quad\quad\quad\quad H_0 h + \tfrac{1}{2}V_0 l = 4M_p;$

(c) $\quad \left[\dfrac{V_0 l}{8}\left(\dfrac{3l}{2h+3l}\right)\right] + 2\left[\dfrac{V_0 l}{8}\left(\dfrac{4h+3l}{2h+3l}\right)\right] + \left[\dfrac{V_0 l}{8}\left(\dfrac{3l}{2h+3l}\right) + \dfrac{H_0 h}{2}\right] = 4M_p,$

or $\quad\quad\quad\quad\quad \tfrac{1}{2}H_0 h + \tfrac{1}{2}V_0 l = 4M_p.$

$$(6.37)$$

It will be seen at once that the beam mechanism, (c), is less critical than the combined mechanism, (b). Equations (6.37) may be displayed on an interaction diagram, fig. 6.8, which shows also the corresponding static collapse equations.

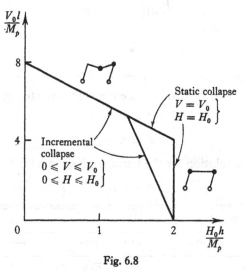

Fig. 6.8

There are two general features illustrated in fig. 6.8 which may be noted. First, the conditions for incremental collapse are always at least as severe as the corresponding conditions for static collapse under the maximum values of the loads; that is, the yield surface for incremental collapse in fig. 6.8 is contained within the yield surface for static collapse. The fact that $\lambda_s \leqslant \lambda_c$ will be discussed more fully in section 6.5 below. Secondly, it is in fact possible for λ_s to equal λ_c, at least for some modes of collapse. By formulating the equations in a different way, as is done in section 6.5, it becomes clear under which conditions there will be a reduction in load factor from λ_c to λ_s.

The equation for mode (b) in (6.37) (which is in fact identical with the corresponding static collapse equation), was written directly from equation (6.28) and the values in table 6.2; the mechanisms (a) and (c) were not combined as suggested in equations (6.31) and (6.32). A numerical example of a fixed-base frame will illustrate the technique of combination of mechanisms. The frame of fig. 6.9 has uniform section of full plastic moment $M_p = 60$ (in arbitrary units); the loads V and H vary independently between the limits

$$\left.\begin{array}{l} 0 \leqslant V \leqslant 30, \\ 0 \leqslant H \leqslant 20. \end{array}\right\} \tag{6.38}$$

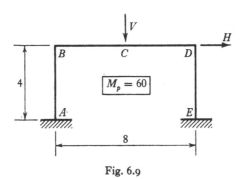

Fig. 6.9

The *static* collapse equations may be laid out as usual, for $V = 30$, $H = 20$; the mechanisms of fig. 6.7 must be taken to have *plastic* hinges at the column feet:

$$\left.\begin{array}{lll} (a) & 80\lambda = 4(60); & \lambda = 3.0, \\ (c) & 120\lambda = 4(60); & \lambda = 2\cdot0, \\ \hline & 200\lambda \quad 8(60) & \\ \text{Cancel hinge } B: & 2(60) & \\ \hline (b) & 200\lambda = 6(60); & \lambda_c = 1\cdot8. \end{array}\right\} \tag{6.39}$$

The combined mode (b) is most critical.

The elastic bending moments in the frame may be determined by any standard method, and table 6.3 gives the results of such an analysis. Thus, writing equation (6.28) in full for mode (a) of fig. 6.7,

$$\lambda[(1)(25)+(-1)(-15)+(1)(39)+(-1)(-37)] = (4)(60),$$

or
$$(a) \quad 116\lambda = 240; \quad \lambda = 2 \cdot 07. \qquad (6.40)$$

Similarly, the beam mechanism, fig. 6.7(c), gives

$$(c) \quad 135\lambda = 240; \quad \lambda = 1 \cdot 78. \qquad (6.41)$$

Table 6.3

Section	Due to V		Due to H		Combined loading		
	$\mathcal{M}^{max}$	$\mathcal{M}^{min}$	$\mathcal{M}^{max}$	$\mathcal{M}^{min}$	$\mathcal{M}^{max}$	$\mathcal{M}^{min}$	$\mathcal{M}^{max} - \mathcal{M}^{min}$
A	0	−12	25	0	25	−12	37
B	24	0	0	−15	24	−15	39
C	0	−36	0	0	0	−36	36
D	24	0	15	0	39	0	39
E	0	−12	0	−25	0	−37	37

These two mechanisms may be combined to give

$$(a) \quad 116\lambda = 240$$
$$(c) \quad 135\lambda = 240$$
$$\overline{ \quad 251\lambda \quad 480}$$

Cancel hinge B:
$$\underline{39\lambda \quad 120}$$
$$(b) \quad 212\lambda = 360; \quad \lambda_s = 1 \cdot 698. \qquad (6.42)$$

The term 39λ subtracted from the left-hand side of the equation to represent the cancellation of hinge B is the value of $(\mathcal{M}_B^{max} - \mathcal{M}_B^{min})(1)$, taken from table 6.3; the hinge rotation at B has been taken as unity in fig. 6.7.

Comparing the values of λ for the three mechanisms represented by equations (6.40), (6.41) and (6.42), it seems that mode (b) is the correct mode of incremental collapse. The only three modes possible have in fact now all been examined, so that it is certain that mode (b) is correct; however, a static check will be made for this mode to illustrate the use of the technique.

The incremental collapse mode involves hinges at A, C, D and E, not, of course, formed all at the same time. The hinge at A has a positive rotation, and forms when the elastic moment is a maximum, that is, when

$$(1\cdot698)(\mathcal{M}_A^{\max}) + m_A = 60, \tag{6.43}$$

where m_A is the value of the residual moment induced at the section. Using the value of $\mathcal{M}_A^{\max}$ from table 6.3,

$$(1\cdot698)(25) + m_A = 60.$$

Similarly
$$\left. \begin{aligned} (1\cdot698)(25) + m_A &= 60. \\ (1\cdot698)(-36) + m_C &= -60, \\ (1\cdot698)(39) + m_F &= 60, \\ (1\cdot698)(-37) + m_E &= -60. \end{aligned} \right\} \tag{6.44}$$

Hence $\quad m_A = 17.5, \quad m_C = 1\cdot1, \quad m_D = -6\cdot2, \quad m_E = 2\cdot9. \tag{6.45}$

Now there are five critical sections in the frame, and three basic redundancies, so that the residual moments at the five sections must satisfy two independent equilibrium equations; mechanisms (a) and (c) of fig. 6.7 give

$$\left. \begin{aligned} m_A - m_B \qquad + m_D - m_E &= 0, \\ m_B - 2m_C \quad + m_D &= 0. \end{aligned} \right\} \tag{6.46}$$

Elimination of m_B from these equations gives

$$m_A \qquad - 2m_C \quad + 2m_D - m_E = 0 \tag{6.47}$$

(which could have been written directly from fig. 6.7(b)). The values (6.45) satisfy equation (6.47) identically, which checks the numerical working; put another way, the value $\lambda_s = 1\cdot698$ could have been deduced from equations (6.44) expressing the formation of the incremental collapse mechanism.

Either of equations (6.46) gives $m_B = 8\cdot4$. Thus, using table 6.3 again, the *actual* value of the bending moment at section B varies between the limits

$$(1\cdot698)(\mathcal{M}_B^{\max}) + m_B = 40\cdot8 + 8\cdot4 = 49\cdot2,$$

and
$$\left. \begin{aligned} (1\cdot698)(\mathcal{M}_B^{\max}) + m_B &= 40\cdot8 + 8\cdot4 = 49\cdot2, \\ (1\cdot698)(\mathcal{M}_B^{\min}) + m_B &= -25\cdot5 + 8\cdot4 = -17\cdot1. \end{aligned} \right\} \tag{6.48}$$

Since both the extreme values are less numerically than the value $M_p = 60$, the yield condition is satisfied at B, and the collapse mechanism is confirmed.

Reference to table 6.3 indicates the combinations of loads which lead to the formation of the hinges. Hinge A forms, for example, when the elastic moment is a maximum (cf. equation (6.43)), that is, when the side load H is applied in the absence of the vertical load V. The other three hinges form when both loads are applied together (the hinge at the centre of the beam is not in fact affected by the side load). Thus the sequence of loading illustrated in fig. 6.10, if repeated at a load factor above 1·698 (and below the static-collapse value of 1·8), would lead to rapid incremental collapse.

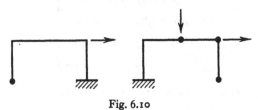

Fig. 6.10

As a second example of shakedown behaviour, the same frame (fig. 6.9) will be analysed on the assumption that the idealized wind load can reverse completely in direction, so that

$$\left.\begin{array}{r} 0 \leqslant V \leqslant 30, \\ -20 \leqslant H \leqslant 20. \end{array}\right\} \tag{6.49}$$

Table 6.4

Section	$\mathscr{M}^{\max}$	$\mathscr{M}^{\min}$	$\mathscr{M}^{\max} - \mathscr{M}^{\min}$
A	25	−37	62
B	39	−15	54
C	0	−36	36
D	39	−15	54
E	25	−37	62

The values of maximum and minimum elastic moments shown in table 6.4 can be deduced immediately from the values of table 6.3. It is found (see example 6.2 below) that the beam mode of collapse is most critical, fig. 6.7(c). The incremental collapse equation is

$$\lambda_s(39+72+39) = 4(60),$$

or

$$\lambda_s = 1·60. \tag{6.50}$$

(Note that in this case there is some danger of alternating plasticity at the column feet A and E, where the (unfactored) range of moment is 62. If the value of the yield moment for the section is $M_y = 50$ (cf. $M_p = 60$), then the load factor against alternating plasticity is $100/62 = 1.61$.)

The statical analysis proceeds as in the previous example:

$$B: \quad (1.60)(39) + m_B = 60, \quad m_B = -2.4,$$
$$C: \quad (1.60)(-36) + m_C = -60, \quad m_C = -2.4, \qquad (6.51)$$
$$D: \quad (1.60)(39) + m_D = 60, \quad m_D = -2.4.$$

These values of residual moments satisfy the second of equations (6.46), but the first gives merely $m_A = m_E$. Since the mechanism of collapse is *partial*, involving three rather than four hinges, the frame remains statically indeterminate at incremental collapse. In order to confirm the collapse mode and the corresponding factor $\lambda_s = 1.60$, a value of m_A must be found to satisfy, at section A, the basic inequalities (6.6). From table 6.4

$$(1.60)(25) + m_A \leqslant 60,$$
and
$$(1.60)(-37) + m_A \geqslant -60. \qquad (6.52)$$

Thus it is certainly possible to find a value $\overline{m}_A$ in the range $20 \geqslant \overline{m}_A \geqslant -0.8$ which will satisfy the shakedown theorem. Although the actual value of m_A cannot be determined from this analysis, the incremental collapse load factor is nevertheless confirmed.

6.5   The relation between λ_c and λ_s

It was mentioned above that λ_s is always less than or at most equal to λ_c, where the value of the static load factor is computed from the maximum values of the loads. In fact, the value of the incremental load factor resulting from the analysis of any *assumed* mechanism of collapse, not necessarily the correct mechanism, can never exceed the corresponding static value of the load factor for the same assumed mechanism. The following formulation was first given by Ogle.†

Suppose a unit load acting at section j of a frame produces an elastic bending moment μ_{ij} at section i of the frame. Then the actual load W_j acting at j will give rise to an elastic bending moment

$$\mathcal{M}_i = \mu_{ij} W_j. \qquad (6.53)$$

In computing the maximum elastic bending moment $\mathcal{M}_i^{\max}$, the value of W_j is chosen to be as large or as small as possible according as μ_{ij} is

† M. H. Ogle, Shakedown of steel frames, Ph.D. Thesis (Cambridge University), 1964.

6. REPEATED LOADING

positive or negative, denoted by μ_{ij}^+ and $-\mu_{ij}^-$, where μ_{ij}^+ and μ_{ij}^- are themselves positive numbers. Thus if

$$W_j^{\min} \leqslant W_j \leqslant W_j^{\max}, \tag{6.54}$$

then
$$\mathscr{M}_i^{\max} = \sum_j (\mu_{ij}^+ W_j^{\max} - \mu_{ij}^- W_j^{\min}),$$

and similarly, $\quad \mathscr{M}_i^{\min} = \sum_j (-\mu_{ij}^- W_j^{\max} + \mu_{ij}^+ W_j^{\min}).$
$$\tag{6.55}$$

The basic incremental collapse equation for any assumed mechanism ϕ_i, equation (6.28), may thus be written

$$\lambda_s[\sum_i\{\sum_j(\mu_{ij}^+ W_j^{\max} - \mu_{ij}^- W_j^{\min})\}\,\phi_i^+ - \sum_i\{\sum_j(-\mu_{ij}^- W_j^{\max} + \mu_{ij}^+ W_j^{\min})\}\,\phi_i^-]$$
$$= \Sigma(M_p)_i\,|\phi_i|. \quad (6.56)$$

The corresponding static collapse equation for the same mechanism is

$$\lambda_c[\sum_i\{\sum_j(\mu_{ij}^+ W_j^{\max} - \mu_{ij}^- W_j^{\max})\}\,\phi_i^+ - \sum_i\{\sum_j(-\mu_{ij}^- W_j^{\max} + \mu_{ij}^+ W_j^{\max})\}\,\phi_i^-]$$
$$= \Sigma(M_p)_i\,|\phi_i|. \quad (6.57)$$

This equation is almost identical with (6.56); the difference is that all loading terms in (6.57) are due to $W_j^{\max}$. Dividing each equation by its load factor, and subtracting,

$$\left(\frac{1}{\lambda_s} - \frac{1}{\lambda_c}\right)\Sigma(M_p)_i\,|\phi_i| = \sum_i \phi_i^+\{\sum_j \mu_{ij}^-(W_j^{\max} - W_j^{\min})\}$$
$$+ \sum_i \phi_i^-\{\sum_j \mu_{ij}^+(W_j^{\max} - W_j^{\min})\}. \quad (6.58)$$

The right-hand side of equation (6.58) is thus a measure of the difference between λ_c and λ_s (or, strictly, between their reciprocals), for any given collapse mechanism ϕ. Note first that it is the *range* of loading $(W_j^{\max} - W_j^{\min})$ only which leads to this difference, and not the absolute values of the loads. As an immediate corollary, therefore, any dead load (or load of fixed magnitude) cannot affect any term in equation (6.58), and may be disregarded in calculating the *difference* between λ_c and λ_s. That is, the dead loads will affect the value of λ_c, but will then play no further part in the analysis if this is done according to equation (6.58).

Secondly, the whole of the right-hand side of equation (6.58) is positive or zero. The range $(W_j^{\max} - W_j^{\min})$ is itself essentially positive or zero, while the products $\phi^+\mu^-$ and $\phi^-\mu^+$ are all, by definition, positive. Thus, from equation (6.58), $\lambda_s \leqslant \lambda_c$ for any mechanism ϕ.

Thirdly, it is *only* products $\phi^+\mu^-$ and $\phi^-\mu^+$ which appear in equation (6.58). The difference between λ_c and λ_s is affected only by those loads for which a unit load would produce a *negative* elastic moment at a section where there is a *positive* hinge rotation, or which would produce a *positive* elastic moment at a section where there is a *negative* hinge rotation.

The features of equations (6.37) and fig. 6.8 are now explicable. Figure 6.11 shows sketch elastic solutions for $H = 1$ and $V = 1$, together with the three modes of collapse. For the sidesway mode (a), for example, the signs of the hinge rotations at B and D are the same as the signs of the

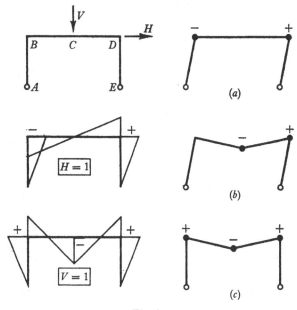

Fig. 6.11

unit bending moments due to $H = 1$. This implies that the static collapse and the incremental collapse equations will contain precisely the same contributions from the side load H, and this may be confirmed by comparing the first of equations (6.35) with the first of (6.37):

$$\text{Static:} \quad Hh = 2M_p,$$
$$\text{Incremental:} \quad H_0 h + \frac{V_0 l}{8}\left(\frac{3l}{2h+3l}\right) = 2M_p. \qquad (6.59)$$

For the same sway mode (a) the vertical load $V = 1$ produces a positive moment at B where the hinge rotation is negative. At D, the moment due to $V = 1$ has the same sign as the hinge rotation of the mechanism. Thus an extra term due to V will arise in the incremental collapse equation compared with the corresponding static equation, *due only to the hinge at B*. By a simple extension of equation (6.58), the contribution is merely the magnitude of the hinge rotation multiplied by the range of *moment* $(\mathcal{M}_B^{\max} - \mathcal{M}_B^{\min})$ which is due to the load V. From table 6.2 this range is

$$\frac{V_0 l}{8}\left(\frac{3l}{2h+3l}\right)$$

and, with unit hinge rotation, this gives the term in the second of (6.59).

Comparing the sketched elastic solutions in fig. 6.11 with the hinge rotations of mode (*b*), *neither* load will contribute to any extra terms to be added to the static collapse equation; the second of (6.35) and (6.37) are identical, and the boundaries of the yield surfaces in fig. 6.8 are the same for this mode. Similarly, examination of the beam mode of collapse indicates that an extra term will arise in the incremental collapse equation from the side load, but not from the vertical load.

From this numerical example, the following general simplification of equation (6.58) is indicated. The right-hand side may be written

$$\sum_i \{\phi_i^+ (\sum_j \mu_{ij}^- \overline{W_j}) + \phi_i^- (\sum_j \mu_{ij}^+ \overline{W_j})\} \equiv \sum_i (\phi_i^+ \overline{\mathcal{M}_i^-} + \phi_i^- \overline{\mathcal{M}_i^+}) \equiv \sum_i |\phi_i \mathcal{M}_i^*|,$$
(6.60)

where $\overline{W_j}$ represents the range of loading $(W_j^{\max} - W_j^{\min})$, leading to a range of elastic bending moment $\overline{\mathcal{M}_j}$, denoted plus or minus according to sign. It is seen that only the products of a *positive* change of moment $\mathcal{M}_i^+$ with a *negative* hinge rotation ϕ_i^-, and vice versa, are taken, and this is indicated by the final short notation $|\phi_i \mathcal{M}_i^*|$ of (6.60). Thus finally, if the static collapse equation is written

$$\text{Static:} \quad \lambda_c\{\sum(M_F)_i \phi_i\} = \sum(M_p)_i |\phi_i|, \tag{6.61}$$

(cf. equation (6.34)), then the incremental collapse equation may be written

$$\text{Incremental:} \quad \lambda_s\{\sum(M_F)_i \phi_i + \sum|\mathcal{M}_i^* \phi_i|\} = \sum(M_p)_i |\phi_i|. \tag{6.62}$$

In these equations, $(M_F)_i$ represents any convenient set of equilibrium bending moments.

As an example, the fixed-base frame of fig. 6.9 and table 6.3 will be reanalysed using equation (6.62). The data of table 6.3 can be rearranged as in table 6.5.

The static collapse equations have been taken from (6.39), and these have then been modified by the $M^*\phi$ terms shown in the table to give the incremental collapse equations. These last are identical with equations (6.40), (6.41) and (6.42).

Table 6.5

Section	Maximum positive and negative change in bending moment		ϕ	$\mathcal{M}^*\phi$	ϕ	$\mathcal{M}^*\phi$	ϕ	$\mathcal{M}^*\phi$
	$\mathcal{M}+$	$\mathcal{M}-$						
A	25	−12	1	12	1	12	—	—
B	24	−15	−1	24	—	—	1	15
C	0	−36	—	—	−2	0	−2	0
D	39	0	1	0	2	0	1	0
E	0	−37	−1	0	−1	0	—	—

Static collapse: $240 = 80\lambda_c$; $360 = 200\lambda_c$; $240 = 120\lambda_c$

Incremental collapse: $240 = 116\lambda_s$; $360 = 212\lambda_s$; $240 = 135\lambda_s$

The first two columns of numbers in table 6.5 are the same as the columns headed $\mathcal{M}^{\mathrm{max}}$ and $\mathcal{M}^{\mathrm{min}}$ for the combined loading shown in table 6.3. This is fortuitous, since for this particular example the minimum values of the loads were zero. Had there been some dead load in addition to the loads V and H, the maximum positive and negative *changes* in moment in table 6.5 would have remained the same, while the numbers in table 6.3 would have altered. That is, the values of λ_c in table 6.5 would have been affected by the presence of dead load, but the $\mathcal{M}^*\phi$ terms would have been the same.

6.6 A two-storey frame

As a final example of this kind of incremental collapse analysis, the two-storey frame of fig. 6.12 will be investigated. (The behaviour of this frame has some similarities to that of the practical design example of chapter 10; the calculations will be given in reasonable detail here.) The loads shown will be taken as varying between zero and their maximum values, and an interaction diagram is to be constructed. The frame has uniform section, full plastic moment M_p.

Of the large number of possible collapse mechanisms, those of fig. 6.13

are the most critical for static collapse of the frame. The static collapse
equations are

$$(a) \qquad \tfrac{1}{2}Vl = 4M_p,$$
$$(b) \quad 2Hl + \tfrac{1}{2}Vl = 8M_p, \tag{6.63}$$
$$(c) \quad \tfrac{3}{2}Hl \qquad\ = 4M_p.$$

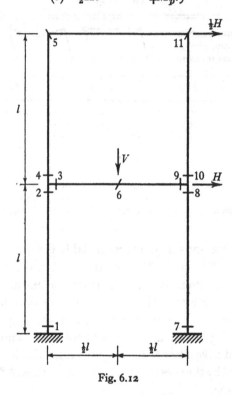

Fig. 6.12

That no other collapse mechanism is critical may be confirmed by making
statical analyses for points A and B of the interaction diagram in fig. 6.14;
it will be found that in both cases equilibrium bending-moment distributions can be constructed which satisfy the yield condition.

The elastic bending-moment distribution in the frame, computed as
usual on the assumption of an initially stress-free structure, is given in
the first two columns of table 6.6. The maximum positive and negative
changes of bending-moment $\overline{\mathcal{M}}$ can then be determined, and the static
collapse equations (6.63) modified by the inclusion of $\mathcal{M}^*\phi$ terms. The
last three sections of table 6.6 give the hinge rotations and the corresponding products $\mathcal{M}^*\phi$ for each of the three modes of fig. 6.13. Although

Table 6.6

Section	Elastic bending moments due to H ($\times Hl\times10^4$)	V ($\times Vl\times10^4$)	Maximum positive and negative change in bending moment $\bar{\mathcal{M}}^+$ ($\times10^4$)	$\bar{\mathcal{M}}^-$ ($\times10^4$)	Mode (a) ϕ	$\mathcal{M}^*\phi$	Mode (b) ϕ	$\mathcal{M}^*\phi$	Mode (c) ϕ	$\mathcal{M}^*\phi$
1	4409	-268	$4409Hl$	$-268Vl$	—	—	1	$0.0268Vl$	1	$0.0268Vl$
2	-3091	536	$536Vl$	$-3091Hl$	—	—	—	—	-1	$0.0536Vl$
3	-3954	981	$981Vl$	$-3954Hl$	1	$0.3954Hl$	—	—	—	—
4	863	-446	$863Hl$	$-446Vl$	—	—	—	—	—	—
5	-1637	89	$89Vl$	$-1637Hl$	—	—	-1	$0.0089Vl$	—	—
6	0	-1519	0	$-1519Vl$	-2	—	-2	—	—	—
7	-4409	-268	0	$-4409Hl-268Vl$	—	—	-1	—	-1	—
8	3091	536	$3091Hl+536Vl$	0	1	—	2	—	—	—
9	3954	981	$3954Hl+981Vl$	0	—	—	—	—	—	—
10	-863	-446	0	$-863Hl-446Vl$	—	—	1	—	—	—
11	1637	89	$1637Hl+89Vl$	0	—	—	1	—	—	—

Table 6.7

Section	$\mathcal{M}^{max}/M_p$	$\mathcal{M}^{min}/M_p$	m/M_p	M^{max}/M_p	M^{min}/M_p	$(\mathcal{M}^{max}-\mathcal{M}^{min})/M_p$
1	1.030	-0.165	-0.030	1.000	-0.195	1.195
2	0.330	-0.722	-0.278	0.052	-1.000	1.052
3	0.603	-0.924	0.397	1.000	-0.527	1.527
4	0.202	-0.274	-0.675	-0.473	-0.949	0.476
5	0.055	-0.383	m_5	—	—	0.438
6	0	-0.934	-0.066	-0.066	-1.000	0.934
7	0	-1.195	0.195	0.195	-1.000	1.195
8	1.052	0	-0.052	1.000	-0.052	1.052
9	1.527	0	-0.527	1.000	-0.527	1.527
10	0	-0.476	0.475	0.475	-0.001	0.476
11	0.437	0	$1.150+m_5$	—	—	0.437

these three modes are critical for static collapse, it does not follow that
they must remain so for incremental collapse; a frame which collapses in
a certain mode under static loading will not necessarily collapse incre-

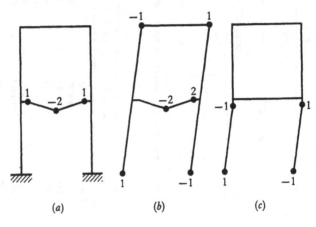

Fig. 6.13

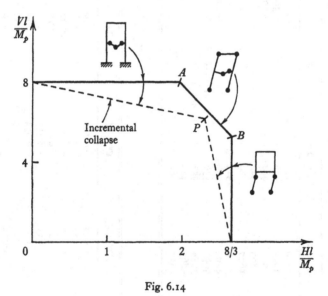

Fig. 6.14

mentally in the same mode when the loads vary independently. (However,
the mode of incremental collapse is likely to correspond to one of those
modes of static collapse which give close upper bounds to the value of λ_c.)

The static collapse equations (6.63) are modified as follows for incremental collapse:

$$
\begin{aligned}
(a)\quad & 0{\cdot}3954Hl + \tfrac{1}{2}Vl = 4M_p, \\
(b)\quad & 2Hl + 0{\cdot}5357Vl = 8M_p, \\
(c)\quad & \tfrac{3}{2}Hl + 0{\cdot}0804Vl = 4M_p.
\end{aligned}
\tag{6.64}
$$

It will be seen that the equation for a mode (b) is only slightly changed, and indeed this mode does not occur with the present range of loading. Instead, the interaction diagram for incremental collapse is defined by the two modes (a) and (c) as sketched in fig. 6.14; a statical analysis will be made for point P to confirm that no other mode is possible.

Point P has coordinates $Hl = 2{\cdot}337M_p$, $Vl = 6{\cdot}151M_p$, and these values lead to the values of $\mathcal{M}^{\max}$ and $\mathcal{M}^{\min}$ given in table 6.7. Now the seven hinges formed simultaneously at P by mechanisms (a) and (c), fig. 6.13, give rise to seven equations for values of residual moments:

$$
\begin{aligned}
1:\quad & 1{\cdot}030M_p + m_1 = M_p; & m_1 = -0{\cdot}030M_p, \\
2:\quad & -0{\cdot}722M_p + m_2 = -M_p; & m_2 = -0{\cdot}278M_p, \\
8:\quad & 1{\cdot}052M_p + m_8 = M_p; & m_8 = -0{\cdot}052M_p, \\
7:\quad & -1{\cdot}195M_p + m_7 = -M_p; & m_7 = 0{\cdot}195M_p, \\
3:\quad & 0{\cdot}603M_p + m_3 = M_p; & m_3 = 0{\cdot}397M_p, \\
6:\quad & -0{\cdot}934M_p + m_6 = -M_p; & m_6 = -0{\cdot}066M_p, \\
9:\quad & 1{\cdot}527M_p + m_9 = M_p; & m_9 = -0{\cdot}527M_p.
\end{aligned}
\tag{6.65}
$$

Since the frame has six redundancies, there must exist five relationships between the eleven values of the residual moments at the critical sections. These five equations may be found in the usual way by considering the sways of the upper and lower storeys, the lower beam mechanism, and joint rotations at the ends of the lower beam:

$$
\begin{aligned}
m_4 - m_5 + m_{11} - m_{10} &= 0, \\
m_1 - m_2 + m_8 - m_7 &= 0, \\
m_3 - 2m_6 + m_9 &= 0, \\
m_2 - m_3 - m_4 &= 0, \\
m_8 - m_9 - m_{10} &= 0.
\end{aligned}
\tag{6.66}
$$

The second and third of equations (6.66) are satisfied identically by the values of (6.65); the virtual mechanisms from which these two equations were derived are the actual incremental collapse mechanisms occurring simultaneously at P, and the arithmetic is thus checked.

The fourth and fifth of (6.66) furnish the values of m_4 and m_{10}, and these are entered in table 6.7 together with the values from (6.65). This leaves the first of (6.66) as a single equation connecting the values of m_5 and m_{11}. The frame of six redundancies is collapsing with seven hinges, but two collapse relationships (i.e. the values of V and H) must be furnished, since two mechanisms are occurring simultaneously; it is to be expected, therefore, that one redundancy will be left.

From table 6.7 it will be seen that the *total* bending moments at sections other than 5 and 11 do not exceed M_p; the seven hinges are formed at the proper cross-sections. Further, any value of m_5 in the range

$$-0 \cdot 587 > m_5 > -0 \cdot 617 \qquad (6.67)$$

will ensure that the total bending moments at sections 5 and 11 do not exceed M_p; point P in fig. 6.14 is thus confirmed as lying on the boundary of the interaction diagram for incremental collapse.

Finally, the last column of table 6.7 gives the range of elastic bending moment at each section. The largest range is $1 \cdot 527 M_p$ for sections 3 and 9 at the ends of the lower beam, and these sections are therefore in some danger of being subjected to alternating plasticity; however, with an I-section of shape factor $1 \cdot 15$ the permitted range is about $1 \cdot 7 M_p$.

6.7 Rolling loads

The repeated passage of a load, or of a train of loads, over a continuous girder may lead to incremental collapse. Although a single application of the given loading system will not produce collapse, it is possible that the same system, if a few passages are made, may again lead to irreversible plastic deformation. This effect is of some practical importance in the design of crane girders, if these are made continuous over two or more spans. (A conventional type of simply-supported girder may suffer from fatigue problems, but, since it is statically determinate, will not give rise to the kind of shakedown behaviour discussed here.)

The problem of the rolling load is straightforward, but numerical computations can be tedious. This is because the *position* of the load is not known *a priori*, but has to be discovered during the course of the analysis. Horne and Davies† discuss the case of two loads at a fixed

† M. R. Horne and J. M. Davies, Repeated loading in the plastic design of crane girders, *The Engineer*, vol. 216, p. 1053, 1963.

distance apart (representing the wheel loads of a crane) acting on a continuous beam system. As a very simple problem, the effect of a single rolling load acting on a propped cantilever will be analysed here.

Figure 6.15(a) shows the propped cantilever with the point load W acting at a distance zl from the simple support; the corresponding *static* collapse mechanism is shown in fig. 6.15(b). This static case will first be

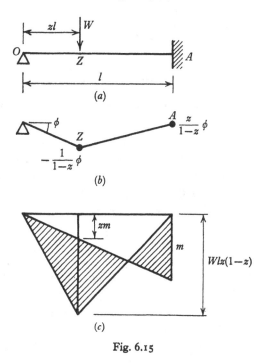

Fig. 6.15

analysed to give a reference value against which the repeated loading case can be measured. If the beam has uniform section, of full plastic moment M_p, then, from fig. 6.15(b),

$$W(zl\phi) = M_p \left(\frac{1+z}{1-z}\right) \phi,$$

or

$$M_p = Wl \frac{z(1-z)}{(1+z)}. \tag{6.68}$$

The most unfavourable position of the load, giving the largest value of M_p, occurs for $z = (\sqrt{2}) - 1)$, and the corresponding value of M_p is

$$M_p = 0 \cdot 686 \left(\frac{Wl}{4}\right). \tag{6.69}$$

Thus, assuming the value of W to incorporate a suitable load factor, a value of M_p given by equation (6.69) is required to prevent collapse of the beam under a *single* passage of the load.

For the shakedown analysis the elastic solution is, as usual, required. The general bending-moment distribution is shown in fig. 6.15(c), and any conventional method of elastic analysis gives

$$m = \tfrac{1}{2}Wlz(1 - z^2).\qquad(6.70)$$

Correspondingly, the elastic bending moment at the section under the load is given, from fig. 6.15(c), by

$$\mathcal{M}_z = -[Wlz(1 - z) - zm] = -\tfrac{1}{2}Wlz(1 - z)^2(2 + z).\qquad(6.71)$$

The mechanism of incremental collapse is that of fig. 6.15(b), and equations (6.70) and (6.71) give the values of the elastic bending moments which must be used in the basic incremental collapse equation (6.28).

Now the hinges at Z and A in fig. 6.15(b) do not form simultaneously at the condition of incremental collapse; that is, the rolling load will be in two different positions as the two hinges form. The maximum elastic moment at the fixed end A will occur when, from equation (6.70), $z = 1/\sqrt{3}$, that is

$$m^{\max} = \frac{Wl}{3\sqrt{3}}.\qquad(6.72)$$

From the bending-moment diagram of fig. 6.15(c) it is clear that the sagging hinge will form under the load, so that, for any given hinge position defined by z, equation (6.71) gives in fact the value of $\mathcal{M}^{\min}$. (It should be noted that the parameter z is being used with two different meanings. In fig. 6.15(a), z defines the position of the *load*, while in fig. 6.15(b) it defines the position of the sagging *hinge* in the incremental collapse mechanism.)

The values of bending moment in (6.71) and (6.72) may now be used with the hinge rotations of fig. 6.15(b) in equation (6.28):

$$[\tfrac{1}{2}Wlz(1 - z)^2(2 + z)]\left[\frac{1}{1 - z}\phi\right] + \left(\frac{Wl}{3\sqrt{3}}\right)\left[\frac{z}{1 - z}\phi\right] = M_p\left(\frac{1 + z}{1 - z}\right)\phi,$$

or

$$M_p = \frac{Wl}{2}\left[\frac{z(1 - z)^2(2 + z)}{(1 + z)} + \frac{2}{3\sqrt{3}}\left(\frac{z}{1 + z}\right)\right].\qquad(6.73)$$

It remains to determine the value of z to give the most critical mechanism of collapse. By differentiation of (6.73)

$$3z^4 + 4z^3 - 3z^2 - 6z + \left(2 + \frac{2}{3\sqrt{3}}\right) = 0,\qquad(6.74)$$

from which, by trial and error,

$$z = 0.372,$$

and
$$M_p = (0.716)\left(\frac{Wl}{4}\right). \tag{6.75}$$

Comparing (6.75) and (6.69), it will be seen that M_p must be increased by some 4 per cent to guard against incremental collapse under repeated passage of the point load.

Such an increase is very small, and for this particular example there is no practical danger of incremental collapse. However, a train of loads acting on a continuous girder might require an increase in full plastic moment of say 30 per cent. In this case, the effect would have to be taken into account in the design, although it may not be necessary to make full allowance for the theoretical increase required. The bare figures for the values of full plastic moment must be examined in the light of some sort of probabilistic calculation before their real significance can be assessed.

6.8   The significance of shakedown calculations

There seems to be general agreement that the danger of collapse of a structure under variable repeated loading is in general less than the danger of collapse under a single application of the maximum values of the loads. To take a specific (but hypothetical) example, suppose that a frame has been designed to collapse under the (static) maximum values of the applied loads, these values having been premultiplied by a specified load factor. For example, a design may have been made using a load factor of 1·40 on dead plus superimposed plus wind loads. The resulting frame is then analysed under the same loading system, but, with the superimposed and wind loads being allowed to vary independently, and it is found that the greatest load factor (defined as in inequalities (6.2)) which can be allowed if shakedown is to occur is say 1.25. The question arises as to whether the static calculations or the shakedown calculations are the more significant in this particular case.

The problem is analogous to that of making a purely static design against wind. As noted in vol. 1, British Standard 449 allows an increase of 25 per cent in *elastic* stresses if such an increase is caused solely by wind. Correspondingly, the implied load factor of 1·75 against collapse under dead and superimposed loading may be reduced to 1·40 when the frame is analysed under dead plus superimposed plus wind loading. Such a decrease in load factor may be thought of as representing the statistical unlikelihood of all loads acting simultaneously.

Thus the numerical problem of shakedown given above seems to resolve itself into assessing the relative probabilities of two events; collapse could occur under a single application of the loads at a load factor λ_c (taken as 1·40 above), or under repeated and random applications at a lower factor λ_s (say 1·25). To take an arbitrary number of cycles, say ten, the question is whether ten repetitions at an overload of 25 per cent are more or less likely to occur than a single overload of 40 per cent.

Put in this way, the problem would seem merely to be one of gathering sufficient statistical information. However, the fact that the arguments can be cast in a probabilistic mould does not mean that they are correct. Certainly there *is* a statistical argument for the case of superimposed and wind loading, but the reduction of load factor from 1·75 to 1·40 (in accordance with the implications of B.S. 449) can be justified only qualitatively from the evidence available. There will in general not be sufficient information to obtain proper frequency curves for the various loads; although ideas of probability distribution (for example) may help in a general assessment of safety, specific calculations will really be pseudo-statistical. Finally, the only argument that can be used is that drawn from experience; the type of frame to which B.S. 449 applies is usually satisfactory on the basis of static load factors of 1·75 and 1·40.

For some types of design problem there is in fact no real statistical argument at all. A certain crane, for example, may be physically incapable of lifting more than a 20 per cent overload. In this case, the question of an *actual* 25 or 40 per cent overload is meaningless. This is not to say that a load factor of 1·40 is not required; the 40 per cent margin will cover not only the 20 per cent overload, but also all those imperfections of theory, calculation and practice which enter inevitably (and, indeed, necessarily) into any engineering design. It is merely that the tail of a normal distribution of loading is simply inapplicable in these circumstances.

There are certain classes of loading for which statistical information *can* help in design. Quite a lot is known, for example, about the distributions of loads on office floors, and the distribution of wind loads has been observed and tabulated. In these cases some meaning can be attached to probability calculations, in the sense that even if there are no errors in fabrication (so that the structure is in perfect accord with the designer's abstract calculations), there is nevertheless a small but finite probability that an overload will occur of magnitude sufficient to cause collapse.

However, as in the case of the crane mentioned above, or in the case of a road bridge designed for the heaviest vehicle allowed on the roads, the statistical arguments finally break down; the tail of the probability curve cannot in practice be correlated with that of an assumed normal distribution. Although calculations are made on the idea of loads being gradually increased by a load factor until collapse occurs, such an increase is, in many cases, purely hypothetical, and cannot be related in any way to an actual overload. The designer thus receives little help with his problem of deciding whether a factor of 1·40 against a hypothetical single overload is of more or less significance than hypothetical repeated loading at a factor of 1·25.

It is important to realize that there is at present no real scientific basis for assessing relative significance in such cases, and experience must ultimately be allowed to decide. Now there appear to be no cases of usual types of building (office blocks, hotels, factory buildings, and so on) in which incremental collapse has been observed. That is, a building designed according to the usual standards, and having implied load factors of 1·75 or 1·40 (depending on whether or not wind is critical in design), has apparently sufficient margin against incremental collapse. Analysis of usual types of frame indicates only a relatively small theoretical drop from λ_c to λ_s, however; for these types of frame, it would seem that a value of λ_s above about 1·25 is adequate to assure safety.

There are unusual types of frame for which a greater drop is possible; such types must be examined individually on their own merits (an example is given in the last chapter of this book). For *usual* types of frame there would seem to be no problem. Despite the concept of load factor, plastic theory is really concerned with the derivation of equilibrium distributions of forces under *working* values of the loads; this point was discussed in the closing pages of vol. 1, and will be returned to in connexion with column design in chapter 9. Plastic theory is usually easier to apply, and makes more reasonable assumptions, than conventional elastic theory, but a final design of a straightforward building will be roughly the same, whichever method is used. It is the actual behaviour of a frame which is in question here, not the design method; if an elastic design is satisfactory from the point of view of incremental collapse, then so will be the (possibly identical) plastic design.

6. REPEATED LOADING

EXAMPLES

6.1 Figure 6.4(a) shows a simply-supported two-span continuous beam of uniform cross-section. The two loads act at mid-span; W_1 can vary in the range $0 \leqslant W_1 \leqslant W$, and W_2 in the range $0 \leqslant W_2 \leqslant kW$, where k is a positive constant less than unity. Show that the value of full plastic moment required to prevent incremental collapse is greater than that needed to prevent static collapse under a single application of the maximum values of the loads by a factor $(1 + \frac{3}{16}k)$.　(*M.S.T.* II, 1967.)

6.2 Investigate all modes of collapse of the frame of fig. 6.9 under the loading (6.49), and verify that the shakedown load factor is given by equation (6.50).

6.3 The members of the fixed-base portal frame shown are of uniform cross-section with full plastic moment M_p, and the welded joints are of full strength. The frame carries a constant vertical load of magnitude $6M_p/l$ at C, and a variable horizontal load H at D.

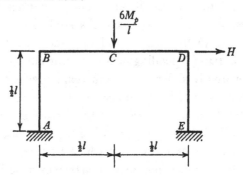

Verify that, according to the simple plastic theory, the value of H to cause static collapse of the frame is $6M_p/l$.

If the side load H now varies between the limits $\pm H_0$, show that the value of H_0 cannot exceed $\frac{16}{3}M_p/l$ if incremental collapse is not to occur. (*M.S.T.* II, 1968.)

6.4 The rectangular portal frame of fig. 6.7 has *fixed* feet and uniform section, full plastic moment M_p. The loads V and H can vary independently within the ranges $0 \leqslant V \leqslant V_0$, $0 \leqslant H \leqslant H_0$. Derive the equations for incremental collapse in the three possible modes.

$$\left(Ans. \; (a) \; Hh + \frac{Vl}{8}\left(\frac{3l}{2l+h}\right) = 4M_p, \right.$$

$$(b) \; Hh + \frac{Vl}{2}\left[1 + \frac{l}{4(2l+h)}\right] = 6M_p,$$

$$\left. (c) \; \frac{Hh}{2}\left(\frac{3h}{l+6h}\right) + \frac{Vl}{2} = 4M_p. \right)$$

6.5 A pinned-base frame of uniform section carries loads V and H as shown in fig. 6.7. Making the usual assumptions of simple plastic theory, an analysis is made to determine the load factor λ_c against collapse under the static loads $V = V_0$, $H = H_0$. A further analysis is made to determine the load factor λ_s against incremental collapse under the random and independent loading $0 \leqslant V \leqslant V_0$, $-H_0 \leqslant H \leqslant H_0$. Show that $\lambda_s = \lambda_c$ if

$$H_0 h < \frac{V_0 l}{4}\left[\frac{1+\frac{4}{3}h/l}{1+\frac{2}{3}h/l}\right].$$
(M.S.T. II, 1965.)

6.6 The loads, V_1, V_2 and H, acting separately and with their maximum static values on the frame shown, give rise to elastic bending moments of values (kN m) shown in the Table. The full plastic moment of the columns is $M_0 = 600$ kN m, and of the beams 1200 kN m. Determine the value of the load factor which would just lead to incremental collapse as the three loads vary randomly and independently between their maximum values and zero.
(*Ans.* 1·49.)

Section	1	2	3	4	5	6	7	8	9	10
V_1	−113	275	262	−156	−35	−7	−802	521	260	113
V_2	−7	−35	−262	156	275	−113	113	260	521	−802
H	207	−160	225	−240	160	−207	−24	113	−113	24

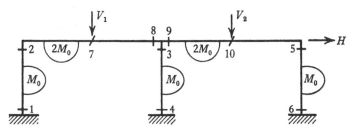

6.7 The continuous beam shown is pinned to rigid foundations; it is of uniform cross-section having full plastic moment M_p, and carries a single concentrated load W which can be placed anywhere on either span. Show that if collapse occurs at a load factor λ_c for a *single* traverse of the load W, then

$$(0\cdot686)\,\lambda_c(Wl/4) = M_p.$$

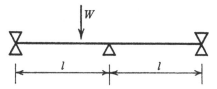

Show further that if incremental collapse is to be avoided for multiple traverses at a load factor λ_s, then

$$(0\cdot700)\,\lambda_s(Wl/4) = M_p.$$
(M.S.T. II, 1966.)

6.8 A fixed-ended beam of uniform section, span l and full plastic moment M_p, is designed just to collapse incrementally under repeated passages of a single point load W. Show that

$$M_p = \tfrac{50}{54}(Wl/8).$$

6.9 A uniform mild-steel beam, encastered at both ends, is to carry a uniformly distributed dead load, together with a concentrated rolling load. The beam is designed by the plastic theory to carry a uniformly distributed load W_D and a central concentrated load W_L with a load factor of 2.0. Determine the maximum value of λ for which a rolling load λW_L, acting in conjunction with a uniformly distributed dead load of λW_D, would just be supported by the beam for any number of crossings of the load.

Sketch a graph of λ against β, where $\beta = W_L/(W_L + W_D)$, for $0 \leqslant \beta \leqslant 1$, and comment on it.

Temperature differences of up to ± 20 deg C may occur between the upper and lower faces of the beam, the temperature gradients being uniform. The temperature gradient at any time does not vary along the length of the beam and is not related in any way to movements of the rolling load. Assuming that the beam is free of temperature stresses when erected, and that the ends are free to move longitudinally, but not to rotate, find the percentage reduction in the load factor due to temperature variations when $\beta = 0.5$. Assume that the beam behaves in a perfectly elastic–plastic manner (shape factor 1.0), and that, for steel, the elastic modulus is $210 \, \text{kN/mm}^2$, the yield stress is $252 \, \text{N/mm}^2$, and the coefficient of expansion is $11 \times 10^{-6} \, \text{deg C}^{-1}$. (*M.S.T.* II, 1964, adapted).

$$\left(Ans. \ \lambda = \frac{54(\beta+1)}{27+32\beta}; \ 100\left(\frac{297}{2017}\right) = 14.7. \right)$$

7

MINIMUM-WEIGHT DESIGN

Most of this chapter will be concerned with the problem of the minimization of the total weight of material in a frame of given geometry acted upon by specified loads. This kind of optimal design is rather restricted. In practice, total material consumption *may* be of prime importance (as in the design of aircraft, for example); more usually for buildings, it is likely that the total *cost* of the structure should be made as small as possible. In addition, other criteria, of a somewhat different kind, may be used to determine specific optimal structures; to take one example, depths of beams might be required to be as small as possible for architectural reasons.

However, weight of material is a parameter which can be handled fairly simply in the structural equations, and at the same time in reasonably general terms; the difficulties are avoided of specifying details of the relative costs of shop and site welding, cost of transport, penalties for the variation of cross-section from member to member, and so on. Further, a minimum weight for a frame provides one absolute base for the designer, who is thus able to compare any practical design against the theoretical minimum.

It is for these reasons that minimum-weight design, rather than optimal design for other criteria, has received most attention in the development of plastic theory. As will be seen, it is possible to formulate general rules for such design, and the programming of such rules for automatic computation is discussed in the next chapter. However, there is a method, that of dynamic programming, for which it is a positive advantage to introduce restraints, such as limitations on the depths of beams; since this method gives an insight into the whole problem of minimum-weight design, while itself being applicable to a much wider range of optimization problems, it will be used to introduce the general topic.

7.1 Dynamic programming

The ideas of dynamic programming, developed by Bellman,[†] will be presented here in a form due to Palmer.[‡] As a first simple example, the

† R. E. Bellman, *Dynamic programming*, Princeton University Press, 1957.
‡ A. C. Palmer, Optimal structure design by dynamic programming, *Proc. Amer. Soc. Civ. Engrs*, **94**, no. ST8, p. 1887, 1968.

propped cantilever of fig. 7.1 (*a*) will be designed; the beam is of uniform cross-section, full plastic moment M_A, and it carries a central point load (whose value is taken to incorporate a suitable load factor). Figure 7.1(*b*) shows the most general bending-moment distribution, with the single redundancy denoted by m_1. From the safe theorem of plastic theory, a possible design of the beam, which will carry the applied load, is one for which the largest bending moment in fig. 7.1 (*b*) does not exceed M_A; a section must be chosen whose full plastic moment is at least equal to the larger of $|240 - \tfrac{1}{2}m_1|$ or $|m_1|$, that is

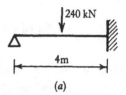

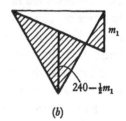

$$M_A = \max\{|240 - \tfrac{1}{2}m_1|, |m_1|\}. \quad (7.1)$$

Working entirely from this statement, and without introducing ideas of mechanisms of collapse, it is clear that equation (7.1) can be examined in a systematic way, by allowing m_1 to vary and recording the corresponding values of M_A. Now there will be a value of m_1 (actually $m_1 = 160\,\mathrm{kN\,m}$ in this example) for which the corresponding value of M_A is least. Thus it is easy to select from equation (7.1), which gives all possible designs of the beam, that one which minimizes the value of M_A.

In varying the value of m_1 it is natural to think of a continuous process; however, the least value of M_A from equation (7.1) does not in fact correspond to a mathematical turning point on some continuous curve. Further, a *practical* design will be derived by examining equation (7.1) for a set of discrete values of m_1, say $m_1 = 0, 10, 20, 30, \ldots$, and finding for each case the corresponding value of M_A. However, instead of using such a regular progression of values for m_1, it is necessary only to consider values which represent the real sections available to the designer. For example, if it is intended that the beam of fig. 7.1 should be constructed using Universal Beam sections, then one of the sections listed in table 7.1 will be used; this table gives the eight most economical light sections that are rolled. The cost of using a section will be taken in this example to be directly proportional to its mass per metre, that is, to the second column of table 7.1; however, any other column of numbers corresponding to each section in the table and representing much more complex cost functions, could be used in the optimization process.

Table 7.2 may now be constructed corresponding to equation (7.1) and fig. 7.1 (b). It is apparent that the 14 × 6¾ UB 45, having $M_p = 193$ kN m, is the lightest section that can be used; the total mass of this section is 180 kg, and this has been entered as the 'cost' X in the last column of table 7.2. It may be noted that the analytical solution

$$M_p = Wl/6 = 160 \text{ kN m}$$

leads to precisely the same design. Table 7.2 represents, however, a *search* of all possible *practical* designs to find by inspection which gives the least cost. In such a process of search, any restriction on the availability of members can only decrease the labour, since the number of discrete values for m_1 in table 7.2 would thereby be reduced.

Table 7.1

Section	Mass/metre, kg	Z_p, mm³	M_p, kN m
8 × 5¼UB	25	259 000	65
8 × 5¼UB	30	313 000	78
10 × 5¾UB	31	395 000	99
12 × 5UB	37	539 000	135
12 × 6½UB	40	623 000	156
14 × 6¾UB	45	772 000	193
14 × 6¾UB	51	893 000	223
15 × 6UB	52	959 000	240

Table 7.2

m_1	Maximum moment in span	M_p used	Cost, X
0	240	240	208
65	208	223	204
78	201	223	204
99	191	193	180
135	173	193	180
156	162	193	180
193	193	193	180
223	223	223	204

The results of table 7.2 may now be used to help in the search for the most economical design of the two-span beam of fig. 7.2 (a). The most general bending-moment diagram is shown in fig. 7.2 (b). If it is required that a beam of constant full plastic moment M_A be used in span A, and a beam of different full plastic moment M_B in span B, the connexion between the two beams being full strength, then a design of the two-span beam will be given by

$$M_A = \max\{|240 - \tfrac{1}{2}m_1|, |m_1|\},$$
$$M_B = \max\{|m_1|, |120 - \tfrac{1}{2}m_1 - \tfrac{1}{2}m_2|, |m_2|\}.$$

(7.2)

(a)

(b)

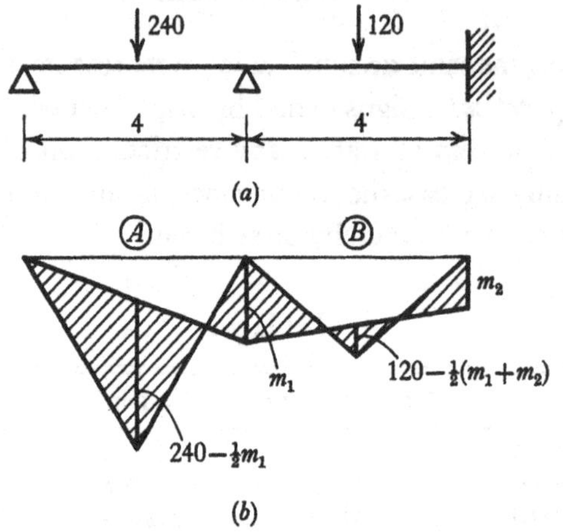

Fig. 7.2

As m_1 and m_2 are varied, a certain combination of M_A and M_B from (7.2) will lead to the design of cheapest cost. In making the search, m_2 will first be put equal to zero, and m_1 varied; the results are given in table 7.3. In this table, the maximum value of bending moment in span B is computed from the expression for M_B in equations (7.2); the value of M_B shown in the table is that of the nearest available section, from which the cost X_B of span B is calculated. *The costs X_A in table 7.3 are reproduced from the previous table 7.2.* As will be seen, this use of previous results is the essence of the present algoristic work.

Table 7.3 ($m_2 = 0$)

m_1	Maximum moment in span B	M_B	X_B	X_A	X_{AB}
0	120	135	148	208	356
65	88	99	124	204	328
78	81	99	124	204	328
99	99	99	124	180	304
135	135	135	148	180	328
156	156	156	160	180	340
193	193	193	180	180	360
223	223	223	204	204	408

The sum of the costs of spans A and B is shown as X_{AB} in table 7.3. The minimum cost is 304 units, and this therefore represents the optimal design of the two-span beam of fig. 7.2 if the right-hand end is simply supported. However, if the right-hand end can carry moment, then the effect of varying the value of m_2 must be investigated. The next step is to construct a table similar to table 7.3, but for $m_2 = 65$; the results are shown in table 7.4.

Table 7.4 ($m_2 = 65$)

m_1	Maximum moment in span B	M_B	X_B	X_A	X_{AB}
0	88	99	124	208	332
65	65	65	100	204	304
78	78	78	120	204	324
99	99	99	124	180	304
135	135	135	148	180	328
156	156	156	160	180	340
193	193	193	180	180	360
223	223	223	204	204	408

Such tables can be constructed for successively increased values of m_2, and the final results for the value of X_{AB} are displayed in table 7.5.

Table 7.5

m_1	Values of X_{AB} for $m_2 =$							
	0	65	78	99	135	156	193	223
0	356	332	328	332	356	368	388	412
65	328	304	324	328	352	364	384	408
78	328	324	324	328	352	364	384	408
99	304	304	304	304	328	340	360	384
135	328	328	328	328	328	340	360	384
156	340	340	340	340	340	340	360	384
193	360	360	360	360	360	360	360	384
223	408	408	408	408	408	408	408	408

For this particular problem, therefore, there would seem to be no advantage in fixing the right-hand end of the beam; the smallest section available is adequate for the right-hand span, whether the end is pinned or fixed. However, table 7.5 can now be used to give, with very little extra labour, the optimal design for the *three*-span beam of fig. 7.3. The work proceeds exactly as before. Starting with $m_3 = 0$, the maximum bending moment in span C can be evaluated for various values of m_2, and hence a design made of span C. For each value of m_2, table 7.5 will then give immediately the minimum cost of X_{AB} of spans A and B.

Table 7.6 ($m_3 = 0$)

m_2	Maximum moment in span C	M_C	X_C	X_{AB}	X_{ABC}
0	240	240	312	304	616
65	208	223	306	304	610
78	201	223	306	304	610
99	191	193	270	304	**574**
135	173	193	270	328	598
156	162	193	270	340	610
193	193	193	270	360	630
223	223	223	306	384	690

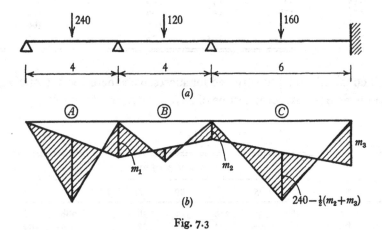

(a)

(b)

$240 - \frac{1}{2}(m_2 + m_3)$

Fig. 7.3

Table 7.7

	Values of X_{ABC} for $m_3 =$							
m_2	0	65	78	99	135	156	193	223
0	616	610	610	574	574	574	**574**	610
65	610	574	574	574	544	**544**	574	610
78	610	574	574	**544**	526	544	574	610
99	**574**	574	**544**	544	526	544	574	610
135	598	**568**	550	550	550	568	594	634
156	610	580	580	580	580	580	610	646
193	630	630	630	630	630	630	630	666
223	690	690	690	690	690	690	690	690

Thus, for this case of $m_3 = 0$, corresponding to a pinned right-hand end of the beam system of fig. 7.3, the work may be laid out as in table 7.6.

Similar tables may be constructed for $m_3 = 65, 78, ...$, and so on, and the complete results are recorded in table 7.7. In the construction of table 7.7, only the *minimum* values of X_{AB}, shown in heavy type in table 7.5, have been used; similarly, the values in heavy type in table 7.7 would be used if the problem were extended to the design of a four-span beam. That is, when considering a particular span of the beam, use is made directly of the results for all previous spans, whose design has already been made optimal.

The *minimum minimorum* in table 7.7 is $X_{ABC} = 526$, occurring for $m_3 = 135$, and $m_2 = 78$ or 99. Working back through table 7.5, it will be seen that the optimum value of X_{AB} for $m_2 = 78$ or 99 occurs for $m_1 = 99$.

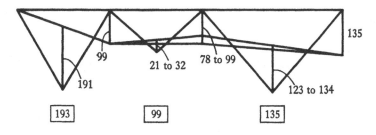

Fig. 7.4

Thus the optimal design of the three-span beam will be given by the bending-moment diagram of fig. 7.4; the corresponding values of M_p required for the three spans are also shown in this figure.

The numerical working of this problem has been given at some length to illustrate the way in which the calculations at any stage make use of the results of all the previous calculations. For a multi-span beam, trial designs of span N are combined with the *known* optimal designs of the previous $(N-1)$ spans; when the design of the N-span beam has been optimized, the problem can be enlarged to one of $(N+1)$ spans, and so on. Since the work *is* entirely numerical, there is no difficulty in introducing complex cost functions to allow, for example, for the processes of fabrication and erection.

The general problem of the multi-span beam may be formulated as follows. In fig. 7.5 is shown such a beam which is supposed to be subjected to loads which are not necessarily either concentrated or uniformly distributed. Whatever the loading, a free bending-moment diagram can be drawn for each span, and a reactant line superimposed; the resulting net diagram can be examined to find the largest bending moment in each span. For span i, the largest bending moment will occur either at the ends, where the bending moments have values m_{i-1} and m_i, or somewhere within the length, where the value will be denoted $\bar{m}_i$. If a uniform section is to be used in the span, then a design can be made if the full plastic moment of span i is equal to at least

$$M_i = \max\{|m_{i-1}|, |\bar{m}_i|, |m_i|\}. \tag{7.3}$$

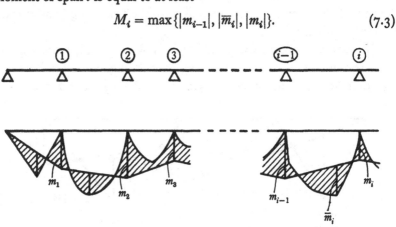

Fig. 7.5

The cost of span i will be a function of certain constants (the length of the span, and of any other parameters that are introduced) and of the value of M_i given by equation (7.3). For given values of the loads, the value of M_i itself is a function only of the reactant moments m_{i-1} and m_i, so that the cost of span i can be expressed as

$$X_i(m_{i-1}, m_i), \tag{7.4}$$

where the function X_i may be as complex as the designer pleases.

If the total cost of the beam can be considered as the sums of the costs of the individual spans, then the total cost of the beam will be

$$X_1(0, m_1) + X_2(m_1, m_2) + \ldots + X_i(m_{i-1}, m_i) + \ldots + X_N(m_{N-1}, m_N). \tag{7.5}$$

The functions X may be different in each span; for the numerical example worked above, the cost was taken merely as the mass in kg of the sections used.

Generalizing and extending the numerical process illustrated by tables 7.2 to 7.7 to the N-span beam, the final stage in the computations would involve the determination of the total cost of the beam system for various discrete values of the *last* reactant moment m_N. For one of these values of m_N, say m, the values of $m_1, m_2, ..., m_{N-1}$ may be varied, and the total cost of the N-span beam recorded; the minimum cost of all such designs will be denoted $C_N(m)$. Thus

$$C_N(m) = \min_{\substack{m_i \\ (i=1,\, 2...,\, N-1)}} \{X_1(0, m_1) + X_2(m_1, m_2) + ... + X_i(m_{i-1}, m_i)$$

$$+ ... + X_N(m_{N-1}, m)\}. \quad (7.6)$$

If the reactant moment m_{N-1} is now picked out from the rest, equation (7.6) can be written

$$C_N(m) = \min_{m_{N-1}} \left\{ \min_{\substack{m_i \\ (i=1,\, 2...,\, N-2)}} \{X_1(0, m_1) + X_2(m_1, m_2) + ... + X_i(m_{i-1}, m_i) \right.$$

$$\left. + ... + X_{N-1}(m_{N-2}, m_{N-1}) + X_N(m_{N-1}, m)\} \right\}. \quad (7.7)$$

The last term of (7.7) can be taken outside the inner minimization process, so that

$$C_N(m) = \min_{m_{N-1}} \left\{ X_N(m_{N-1}, m) + \min_{\substack{m_i \\ (i=1,\, 2...,\, N-2)}} \{X_1(0, m_1) + X_2(m_1, m_2) \right.$$

$$\left. + ... + X_i(m_{i-1}, m_i) + ... + X_{N-1}(m_{N-2}, m_{N-1})\} \right\}, \quad (7.8)$$

and, finally,

$$C_N(m) = \min_{m_{N-1}} \{X_N(m_{N-1}, m) + C_{N-1}(m_{N-1})\}. \quad (7.9)$$

Equation (7.9) is the algorism that was used in working the three-span beam example; the cost of the single-span beam of fig. 7.1 may be written

$$C_1(m) = X_1(0, m) \quad \text{(table 7.2)}, \quad (7.10)$$

and, progressively,

$$C_2(m) = \min_{m_1} \{X_2(m_1, m) + C_1(m_1)\} \quad \text{(table 7.5)}, \quad (7.11)$$

$$C_3(m) = \min_{m_2} \{X_3(m_2, m) + C_2(m_2)\} \quad \text{(table 7.7)}. \quad (7.12)$$

In any given span i of the continuous beam only two variables (m_{i-1} and m_i) were involved in the search process, and the minimum cost of the first r spans could be expressed as a functional $C_r(m_r)$ of a single variable m_r. In extending the ideas of dynamic programming to the general problem of a multi-storey multi-bay frame, a single variable no longer suffices; indeed, $(M+1)$ variables are required for a frame of M bays, which number is, however, independent of the number of storeys. This enlargement of the number of variables, while introducing no difficulties into the theory, makes the search problem much more tedious.

The method is, in fact, efficient for problems such as that of the continuous beam, and it has the immense merit that complicated cost functions can be used with ease. In addition, attention is confined to the known properties of real sections, and the more limited the choice, for architectural or any other reasons, the quicker is the work of the method. The main disadvantage is that the computations are entirely numerical, and no general principles of design emerge from the solution of a particular problem. Such general principles can be discovered if an apparently drastic simplification is made to the cost function, and most of the rest of this chapter will discuss minimum *weight* design under the assumption of a linearized weight function.

7.2 The linear weight function

The sections in table 7.1 (and heavier sections not shown) have masses w per unit length which are given closely by

$$w = aM_p^n, \tag{7.13}$$

where a is a constant and n is approximately o·6. Instead of the discrete number of sections shown in table 7.1, it will be assumed that there is a continuous range of sections available, so that any required full plastic moment can be exactly matched by an actual section. The weights per unit length, and also the unit costs, will be taken to be given by equation (7.13).

The two-span beam of fig. 7.6 will be designed on the assumption that a beam of uniform section, full plastic moment M_A, is used for span A, and a beam of full plastic moment M_B for span B (cf. fig. 7.2; the discrete solution for this problem was given in table 7.3: $M_A = 193$, $M_B = 99$). The bending-moment diagram sketched in fig. 7.6 represents the most general state of bending moments; with the general proportions of this figure (i.e. with $160 > m > 80$), a possible design of the system is

$$\left. \begin{aligned} M_A &= 240 - \tfrac{1}{2}m, \\ M_B &= m. \end{aligned} \right\} \tag{7.14}$$

Such a design would be just on the point of collapse in the mode sketched in fig. 7.7.

Using equation (7.13), the weight X of this design is, to some scale,

$$X = (240 - \tfrac{1}{2}m)^n + (m)^n. \tag{7.15}$$

It is easy to show that, for $0 < n \leqslant 1$ (and in particular for $n \sim 0.6$), X has no minimum value in the range $160 > m > 80$, instead, dX/dm is always positive. Thus to reduce the value of the weight X, the value of m

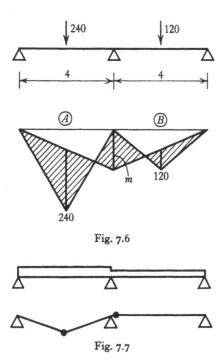

Fig. 7.6

Fig. 7.7

should be reduced. There is, however, a physical limitation to the value of m; at $m = 80$, a second hinge forms in span B, and a smaller section cannot be used for this span. Thus the minimum-weight design of the two-span beam, using the continuous formula (7.13), is given from equations (7.14) by

$$\left. \begin{aligned} M_A &= 200, \\ M_B &= 80. \end{aligned} \right\} \tag{7.16}$$

It may be noted that this solution is not far removed from the discrete solution of table 7.3, $M_A = 193$, $M_B = 99$. Further, the design given by (7.16) was derived independently of the precise value of the index n in the range $0 < n \leqslant 1$; for example, for $n = 1$, equation (7.15) shows at once that $X = 240 + \frac{1}{2}m$, so that, to reduce X as much as possible, m must be given its minimum value of 80.

In the following work, therefore, a *linearized* weight function will be used. If a prismatic member of length l_i is used in span i, having full plastic moment $(M_p)_i$, then the weight of the frame will be taken to be given to some scale by

$$X = \Sigma l_i (M_p)_i. \tag{7.17}$$

In this context, a 'span' is taken as the sum of all separate spans which may be constrained to have the same section. For example, the two columns of a rectangular portal frame might have the same cross-section, while the beam might have a different full plastic moment; the columns together would then form one span, and the beam another.

The general problem, then, is to minimize the value of X given by equation (7.17) for a given frame acted upon by given loads, subject to the constraints imposed by the simple plastic theory of frames. These constraints are simply that any solution must satisfy the three conditions of mechanism, equilibrium and yield; as usual, it is not necessary to write equations which satisfy all three conditions simultaneously.

For example, an approach may be made by satisfying equilibrium and yield alone. Using the problem of fig. 7.6 as an example, the construction of the bending-moment diagram, as a proper combination of free and reactant distributions, ensures that equilibrium is satisfied. The yield conditions may be written for each of the three critical sections:

$$\left.\begin{aligned}
-M_A &\leqslant 240 - \tfrac{1}{2}m \leqslant M_A, \\
-M_A &\leqslant \quad\;\; m \leqslant M_A, \\
-M_B &\leqslant \quad\;\; m \leqslant M_B, \\
-M_B &\leqslant 120 - \tfrac{1}{2}m \leqslant M_B.
\end{aligned}\right\} \tag{7.18}$$

The double continued inequality must be written for the central support since it is not known *a priori* whether $M_A \lessgtr M_B$.

The minimum-weight problem is to minimize the expression

$$X = M_A + M_B \tag{7.19}$$

subject to the restraint of inequalities (7.18). This is a standard problem of linear programming, which is mentioned again briefly in the next chapter. It may be shown that, despite the fact that the mechanism condition has not been mentioned, (7.18) and (7.19) contain enough information for the solution of the problem. Indeed, the direct solution of inequalities (7.18) leads, by one method, to the generation of all possible mechanisms of collapse; it is more fruitful at the moment to discuss the problem of minimum-weight design in terms of the direct construction of mechanisms.

Figure 7.8 illustrates the only four possibilities for collapse of the two-span beam of fig. 7.6; collapse must occur in one span or the other, and the hinge at the internal support will form in span A, or span B, depending

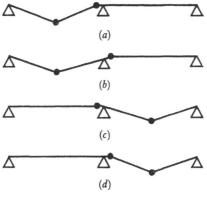

Fig. 7.8

on the relative magnitudes of M_A and M_B. The four collapse equations corresponding to the four mechanisms are

$$
\begin{aligned}
(a) \quad & 3M_A && = 480, \\
(b) \quad & 2M_A + M_B && = 480, \\
(c) \quad & M_A + 2M_B && = 240, \\
(d) \quad & 3M_B && = 240.
\end{aligned}
\qquad (7.20)
$$

These four equations can be plotted in the *design plane* of fig. 7.9. Equation (*a*), for example, plots as the line $M_A = 160$. A design of the frame for which $M_A > 160$ is one which cannot collapse by this mechanism. Similarly, a design for which $2M_A + M_B > 480$ cannot collapse by mode (*b*). Equations (7.20), in fact, define an open *permissible region*, convex when viewed from the origin, in which all safe designs must lie. A point on the boundary of the permissible region represents a design of the frame which is just collapsing, the particular mode of collapse depending on the part of the boundary on which the point lies. A point on the origin side of the permissible region represents a design of the frame which cannot carry the given loads; in this particular numerical example, mode (c) cannot occur at all.

Permissible designs must now be examined to find which is of least weight. For a *given* weight of frame, say $X = 600$, all permitted designs of the two-span beam system will lie on the line PQ in fig. 7.9; this line is simply the plot of the weight function (7.19). Designs of lighter weight

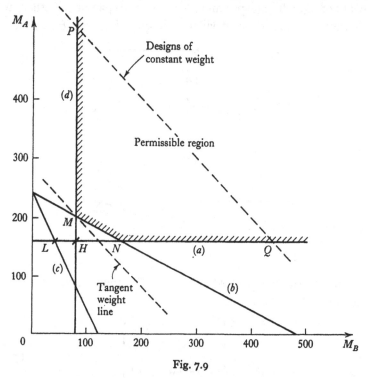

Fig. 7.9

than 600 units will lie on a line parallel to PQ, but closer to the origin, and the minimum weight will occur when the weight line, equation (7.19), is just *tangent* to the permissible region. For the numerical example of fig. 7.9, the point of tangency is the apex M, whose coordinates are given by (7.16).

The tangent weight line in fig. 7.9 lies in the acute angle formed by mechanisms (b) and (d). In analytical terms, equations (b) and (d) of (7.20) can be combined with *positive* multipliers (say α and β) to give the weight equation. Thus,

$$2\alpha M_A + \alpha M_B = 480\alpha,$$

$$3\beta M_B = 240\beta,$$

$$\overline{2\alpha M_A + (\alpha + 3\beta)\,M_B = (480\alpha + 240\beta).} \tag{7.21}$$

If α is taken as $\frac{1}{2}$, and β as $\frac{1}{8}$, then the coefficients of M_A and M_B in equation (7.21) are both unity, and therefore match those of equation (7.19); comparing these two equations, it is seen that the minimum weight is given by $X = 280$.

These ideas may be extended to the design of general frames of this type, and this is done in section 7.3 below. Before moving on, however,

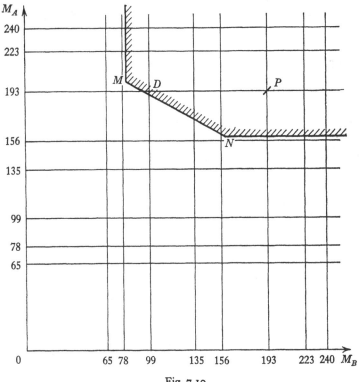

Fig. 7.10

it is of interest to illustrate in the design plane both the boundary of the permissible region and the designs that are possible using the actual discrete sections of table 7.1. The grid in fig. 7.10 corresponds to the eight sections listed, and each grid point corresponds therefore to a possible practical design. If the grid point (such as P) lies within the permissible region then the design will be satisfactory, in that it will carry the given loads. The grid point D ($M_A = 193$, $M_B = 99$) is close to the apex M of the permissible region, and it was this point that was found by the search method of section 7.1.

Figure 7.9 may also be used to illustrate the effect of assuming a linearized weight function. If, instead of the linear equation (7.19), the more realistic function

$$X = (M_A)^{0\cdot6} + (M_B)^{0\cdot6} \qquad (7.22)$$

had been used (which, however, still assumes that a continuous range of sections is available), then the same minimum-weight design would have been derived; the arguments were based, it will be remembered, on an examination of equation (7.15). The form of the weight function does

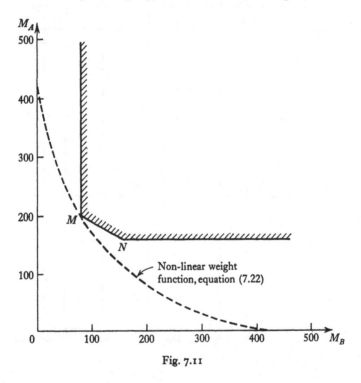

Fig. 7.11

not, of course, affect the mechanism equations (7.20); the idea of a permissible region is unchanged, but the tangent weight line of fig. 7.9 becomes the tangent weight *curve* of fig. 7.11. It will be seen that the curve touches the permissible region at the same apex M. However, had the slope of the line MN been close to that of the weight function in the same region, it is possible that apex N might have given the minimum-weight design, while the linearized theory continued to select apex M. In such a case, however, the two weights represented by M and N would have differed by very little. In any practical design, the use of an 'exact' weight function of the type (7.22) will lead to a design which is either identical with that of the linear theory, or which differs by an insignificant amount.

7.3 Foulkes's theorem

The idea of a permissible region, and the general conditions for minimum weight using a linearized weight function, were first established by Foulkes.† The two-dimensional permissible region (illustrated in fig. 7.9) is extended to N dimensions, for the general frame having N independent full plastic moments whose values have to be assigned in order to achieve minimum weight. Mathematically, the problem is to minimize the linearized weight function X given by equation (7.17) subject to restraining inequalities of the type

$$\Sigma(M_p)_i |\phi_i| \geqslant \Sigma W_j \delta_j. \tag{7.23}$$

The mechanisms (δ, ϕ) of inequalities (7.23) must be a complete set; that is, they must include all possible mechanisms of collapse. The left-hand side of (7.23) will, for a given mechanism, be a linear function of the full plastic moments $(M_p)_i$, and the right-hand side will be a constant for given loads. Thus this particular equation will, if plotted in an N-dimensional design space with axes $(M_p)_i$, be a hyperplane dividing the space into a permitted and a prohibited region. The hyperplanes ('flats') corresponding to all possible mechanisms of collapse will define a permissible region, which will be bounded by the most 'critical' of these hyperplanes. The two-dimensional permissible region was illustrated numerically in fig. 7.9, being bounded by the most critical of (7.20). A problem involving three full plastic moments can be imagined in three dimensions, with the faces of the permissible region being planes, intersecting in lines, the lines themselves intersecting in vertices; each vertex is characterized by the intersection of three planes.

In N dimensions, a vertex of the permissible region is characterized by the intersection of N mechanism flats. Since the coefficients of the left-hand side of (7.23) are essentially positive, the boundary of the permissible region is convex towards the origin, and the region itself is of unlimited extent. The minimum-weight problem consists in the examination of the designs falling within the permissible region.

† J. D. Foulkes, The minimum-weight design of structural frames, *Proc. Roy. Soc.* A, vol. 223, 1954.

Now the weight function (7.17) is itself a flat in N dimensions. For a given value of the weight X, the weight flat will in general intersect the permissible region (as line PQ in fig. 7.9). As the weight is reduced, so the weight flat moves towards the origin, and the minimum will occur when the weight flat is just tangent to the permissible region. At this minimum condition, the weight flat will pass through a vertex, and, stated in these terms, the minimum weight may be found by an examination only of the vertices of the permissible region. (It may be, by numerical accident, that a flat of the boundary of the permissible region is parallel to the weight flat. In that case, any design lying on that sector of the permissible region will be of minimum weight, but the tangent weight flat will still, in fact, pass through a vertex. The reader may care to check that the two-span beam of fig. 7.12 gives the same permissible region shown in fig. 7.9, but that the tangent weight line in this case lies exactly along MN; all designs on MN have the same weight.)

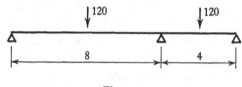

Fig. 7.12

The minimum-weight vertex of the permissible region may be located by writing the condition of tangency of the weight flat. This condition is that the equations of the mechanism flats defining the vertex can be combined with positive multipliers (or, strictly, non-negative multipliers) to give the equation of the weight flat. Now the coefficient of any one full plastic moment, $(M_p)_k$ say, in inequality (7.23) is simply the sum of all the hinge rotations occurring in span k for the particular mechanism (δ, ϕ) being considered. Thus an alternative formulation of the condition of tangency is that the hinge rotations in an individual span must sum (to some scale) to the length of that span.

For a frame composed of prismatic members, therefore, divided into a number N spans with different full plastic moments, the minimum-weight design will be one which is collapsing by the formation of (at least) N simultaneous mechanisms. Further, this *Foulkes mechanism* has hinge rotations which accord in sign with those of the corresponding full plastic moments, and which sum numerically for any particular span to the length of that span (to some arbitrary scale). Finally, the bending-moment distribution corresponding to Foulkes's mechanism must be such that the yield condition is satisfied throughout the frame. (This yield condition is necessary to exclude 'pseudo' vertices such as L or H in fig. 7.9. The geometrical conditions are satisfied at L, but the point does not lie on the boundary of the permissible region.)

The above paragraph, which may be called Foulkes's theorem, will first be illustrated by the general case of the two-span beam. In fig. 7.13 it is supposed that a uniform full plastic moment M_A is provided for the left-hand span, and M_B for the right-hand span; the weight of the beam is therefore

$$X = l_A M_A + l_B M_B. \tag{7.24}$$

Values of M_A and M_B must be chosen to minimize X, subject to the condition that the beam is to carry the fixed central point loads W_A and W_B (whose values incorporate a suitable load factor).

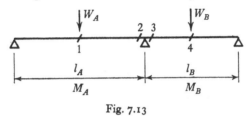

Fig. 7.13

Since there are two design variables, Foulkes's theorem requires the simultaneous formation of two mechanisms. Now fig. 7.8 shows the complete set of all possible mechanisms for the two-span beam which have one degree of freedom; Foulkes's mechanism must therefore be composed of two of the four mechanisms sketched in fig. 7.8. Mechanisms (a) and (b) can be combined as in (i) of fig. 7.14; from the equilibrium of the central joint, this mechanism is possible only for $M_A = M_B$. Similarly, (a) and (c) in fig. 7.8 can be combined to give (ii) of fig. 7.14; however (a) and (d) cannot be combined without violating equilibrium, since the condition $M_A = M_B$ is required at the central support, whereas $M_A = \frac{1}{6}W_A l_A$ from (a) and $M_B = \frac{1}{6}W_B l_B$ from (b). The mechanisms sketched in (iii) and (iv) in fig. 7.14 are also possible, but no further combinations can be constructed.

Mechanism (i) of fig. 7.14 will now be examined to find the conditions under which it represents a Foulkes mechanism. From the conditions of equilibrium and yield, it is seen first that

$$M_A = M_B = \frac{1}{6}(W_A l_A), \tag{7.25}$$

185

and that, to prevent the formation of a hinge at the centre of the right-hand span

$$W_B l_B \leqslant W_A l_A. \tag{7.26}$$

The hinge rotations in fig. 7.14(i) are computed in terms of two parameters θ_1 and θ_2, themselves positive to give bending moments of the right sign at sections 1 and 3 (cf. fig. 7.13). If the hinge rotation at section 2

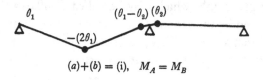

$$(a)+(b) = \text{(i)}, \quad M_A = M_B$$

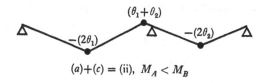

$$(a)+(c) = \text{(ii)}, \quad M_A < M_B$$

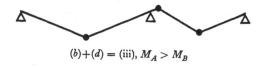

$$(b)+(d) = \text{(iii)}, \quad M_A > M_B$$

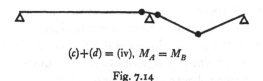

$$(c)+(d) = \text{(iv)}, \quad M_A = M_B$$

Fig. 7.14

is also to be correct, then $\theta_1 \geqslant \theta_2$. Now Foulkes's theorem requires that the numerical sum of the hinge rotations in span A should equal l_A, that is

$$\left.\begin{aligned} 3\theta_1 - \theta_2 &= l_A, \\ \theta_2 &= l_B. \end{aligned}\right\} \tag{7.27}$$

and, similarly,

Thus $\theta_1 = \frac{1}{3}(l_A + l_B)$ and if $\theta_1 \geqslant \theta_2$, then $\frac{1}{3}(l_A + l_B) \geqslant l_B$, or

$$l_A \geqslant 2l_B. \tag{7.28}$$

Mechanism (i) of fig. 7.14 will occur only if the two conditions (7.26) and (7.28) are satisfied; these conditions are shown graphically in fig. 7.15.

Similarly, mechanism (iii) of fig. 7.14, which occurs for $M_A > M_B$, is subject to the conditions

$$\frac{W_A l_A}{W_B l_B} \geqslant 1, \quad \frac{l_A}{l_B} \leqslant 2, \tag{7.29}$$

and the remaining two possibilities are also sketched in fig. 7.15. This plot is complete; note that the numerical problem of fig. 7.7, $l_A/l_B = 1$, $W_A l_A/W_B l_B = 2$ falls in region (iii) with $M_A > M_B$.

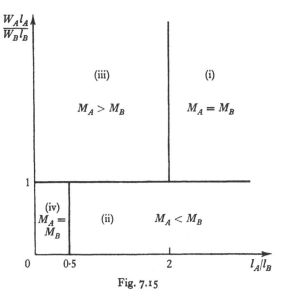

Fig. 7.15

The simultaneous mechanisms (i) and (iv) involve an elementary joint mechanism at the internal support, and are characterized by $M_A = M_B$. It is of course of some practical significance that the minimum-weight design may be one involving a frame of uniform cross-section.

7.4 Limitations on sections

The rectangular portal frame of fig. 7.16 will be designed for minimum weight. The two columns will not be taken as necessarily of equal section, so that there are three full plastic moments to be found, M_A, M_B and M_C. The weight of the frame is given by

$$X = 5M_A + 8M_B + 5M_C. \tag{7.30}$$

Foulkes's mechanism must be of three degrees of freedom, and one or two trials show that a possible minimum-weight design, which must, however, be tested by Foulkes's theorem, would collapse in the mechanism of fig. 7.17. Equilibrium considerations lead quickly to

$$M_A = M_B = 160, \quad M_C = 120, \quad X = 2680, \qquad (7.31)$$

but this is not, in fact, the minimum weight. The hinge rotations of fig. 7.17 must be such that the quantities in parentheses are all positive,

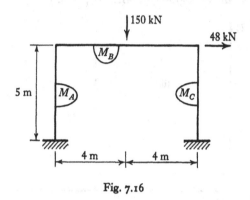

Fig. 7.16

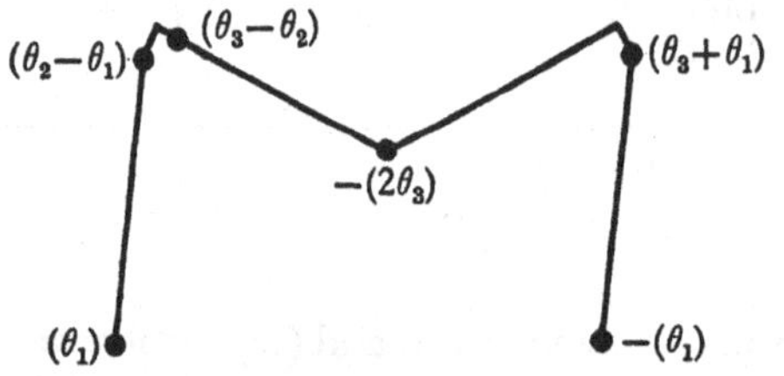

Fig. 7.17

in order to accord with the signs of the full plastic moments. The sums of the rotations in each span give

$$
\left.\begin{aligned}
\theta_2 &= 5, \\
-\theta_2 + 3\theta_3 &= 8, \\
2\theta_1 + \theta_3 &= 5,
\end{aligned}\right\}
\quad \text{or} \quad
\left.\begin{aligned}
\theta_1 &= \tfrac{1}{3}, \\
\theta_2 &= 5, \\
\theta_3 &= 4\tfrac{1}{3}.
\end{aligned}\right\}
\qquad (7.32)
$$

It will be seen that the condition $\theta_3 > \theta_2$ is not satisfied, so that fig. 7.17 does not represent a true Foulkes mechanism.

Figure 7.17 does, however, correspond to a vertex of the permissible region, and it is certain that a neighbouring vertex, while not necessarily that of minimum weight, will nevertheless represent a lighter design. The difficulty is to know in which direction to move, since, equally, some of the neighbouring vertices could give a greater weight. This question is discussed briefly in section 7.5 below and in chapter 8 on automatic computational methods; it turns out that if one of the 'offending' hinges in a trial Foulkes mechanism is suppressed, then an immediate indication is obtained of the way the permissible region should be explored.

There is in fact only one hinge in fig. 7.17 whose rotation is not compatible with the sign of the bending moment, and fig. 7.18 shows the mechanism of *two* degrees of freedom which results from the suppression of this particular hinge. Such a mechanism for a three-dimensional permissible region represents an *edge* of the region, and the design point will be moved along this edge in order to lower the total weight. The virtual work equation resulting from the mechanism of fig. 7.18 and the loads of fig. 7.16 gives

$$(M_A)(\theta_3)+(M_B)(2\theta_3)+(M_C)(2\theta_1+\theta_3) = 240\theta_1+600\theta_3. \quad (7.33)$$

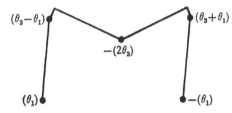

Fig. 7.18

Equation (7.33) holds for any values of θ_1 and θ_3, and hence gives two relationships between the three full plastic moments; for example setting $\theta_1 = 0$ and $\theta_3 = 0$ in turn leads to

$$\left.\begin{aligned} M_A+2M_B &= 480, \\ M_C &= 120. \end{aligned}\right\} \quad (7.34)$$

Thus, from equation (7.30)

$$X = 5M_A+8M_B+5M_C = 2520+M_A. \quad (7.35)$$

The value of M_A must be made as small as possible in order to reduce the value of X. Now there is no impediment, from the point of view of the yield condition or of equilibrium, to reducing the value of M_A to zero, and this is in fact the minimum-weight design:

$$M_A = 0, \quad M_B = 240, \quad M_C = 120, \quad X = 2520. \quad (7.36)$$

For a practical design there will be some minimum value of M_A that can be used, and such *limitations* can be introduced very easily into the analysis. For example, if the actual column used in the design must have, for reasons of stability, a minimum full plastic moment of 60, then equations (7.34) and (7.35) lead to the values

$$M_A = 60, \quad M_B = 210, \quad M_C = 120, \quad X = 2580. \qquad (7.37)$$

(A full plastic moment of about 60 kN m represents the lightest 8×8 UC section; the slenderness ratio of the columns in fig. 7.16 would then have a maximum value of about 100.)

If such limitations are introduced, Foulkes's theorem has to be modified slightly. The design represented by equations (7.37) is collapsing in the mode of *two* degrees of freedom of fig. 7.18, rather than the three required

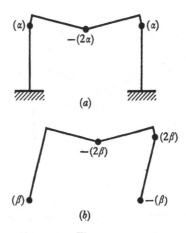

Fig. 7.19

for a Foulkes mechanism. However, the equation representing the limitation on the section of the column is of exactly the same form as a mechanism equation, being a particular case of a linear relation between full plastic moments. Thus the mechanism of fig. 7.18 may be regarded as being built up from the two mechanisms of fig. 7.19 (note that the pure sway mechanism cannot be a constituent of fig. 7.18, since the sign of the bending moment at the top of the left-hand column would be wrong). Figure 7.19 gives the two linear relations:

$$\begin{rcases} \alpha M_A + 2\alpha M_B + \alpha M_C = 600\alpha, \\ \beta M_A + 2\beta M_B + 3\beta M_C = 840\beta, \\ \gamma M_A \qquad\qquad\qquad = 60\gamma. \end{rcases} \qquad (7.38)$$

to which may be added

Equations (7.38) may be solved to give the values (7.37). Further, if they can be combined with positive values of α, β and γ to give the weight equation (7.30), then Foulkes's theorem will be satisfied. It will be seen that this can be done for $\alpha = 3\frac{1}{2}$, $\beta = \frac{1}{2}$, $\gamma = 1$, and that the right-hand sides of equations (7.38) then sum to $X = 2580$.

7.5 Upper and lower bounds

The exploration of the borders of the permissible region to discover the minimum-weight vertex can be a lengthy process, at least if the computations (for a relatively small frame) are done by hand. The problem here is analogous to that of the collapse analysis of a *given* frame. There will be a large number of possible mechanisms, and thus a large number of possible vertices to be examined. Fortunately it is fairly simple to establish bounds to the minimum weight of a frame; if these bounds are close, there may be little point in refining a particular design and, in any case, the method to be used often indicates the way in which modifications should be made.

An upper bound. Any design of a frame which will carry the given loads gives, of course, an upper bound to the minimum weight. To be more precise, a frame which is on the point of collapse, and for which a bending-moment distribution can be constructed which satisfies the yield condition, will be a design lying on the boundary of the permissible region. (Naturally, a point lying *within* the permissible region is also an upper bound.) One of the best ways of finding a design of a particular frame is, in fact, to use the results of a lower-bound analysis.

A lower bound. A design which is collapsing in a Foulkes mechanism (with at least N degrees of freedom), but for which a bending-moment distribution satisfying the yield condition has not been constructed, is a lower bound to the minimum-weight design. As has been mentioned, a point such as L or H in fig. 7.9 represents such a lower bound. (If a proper bending-moment distribution *can* be constructed, then the design is the minimum-weight design.)

The point H in fig. 7.9 is of some significance as a starting-point for practical calculations. It is given by the intersection of the mechanism lines (a) and (d), these corresponding to the failure, in isolation, of either span of the beam, fig. 7.8. The values $M_A = 160$ and $M_B = 80$ are in fact the physical minima that can be used in the design; in a plot such as that of fig. 7.9, the point H represents a square corner. In N dimensions, the analogous point H would represent the corner of a hypercube and, as such, will always correspond to a Foulkes mechanism, for which, however, a satisfactory set of bending moments cannot necessarily be constructed. Indeed, if H is defined by analogy with fig. 7.9, it will always lie *outside* the permissible region.

However, if 'artificial' constraints on the values of the full plastic moments are introduced (as in section 7.4 above) to allow for stability, or for any other reason, then it is possible that the point H will lie within the permissible region, in which case it will represent the actual minimum-weight design. Thus, in fig. 7.10, if it is specified that both M_A and M_B should exceed $193\,\mathrm{kN}\,\mathrm{m}$, then point P must be the design of minimum weight.

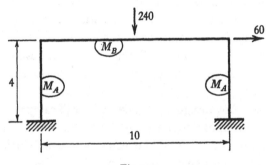

Fig. 7.20

A numerical example will illustrate the technique of upper and lower bounds. The frame of fig. 7.20 carries the loads shown; the two columns are constrained to have the same uniform section of full plastic moment M_A, and the beam has a different full plastic moment M_B. The weight of the frame is given by

$$X = 4M_A + 5M_B. \tag{7.39}$$

If the beam is thought of for the moment as being very strong, then collapse will occur by the pure sideway mode, for which $M_A = 60$. Similarly, if the columns are very strong, failure will occur by the beam mode, for which $M_B = 300$. These two values are evidently the minimum possible that can be used, and correspond to a point 'H' as discussed above. The design is thus a lower bound to $X_{\min}$, so that

$$X_{\min} \geqslant 1740. \tag{7.40}$$

Now a design having $M_A = 60$, $M_B = 300$ will not carry the given loads. A simple analysis shows that collapse will occur by the mode of fig. 7.21 at a load factor of 7/12. If, therefore, the design is strengthened throughout in the ratio 12/7, that is, if M_A is increased to 103 and M_B to 514, a design will result (of weight 2983) which will just carry the given loads, so that

$$2983 \geqslant X_{\min} \geqslant 1740. \qquad (7.41)$$

The lower bound is poor, but the upper bound is in fact only some 10 per cent high; the main use of the point H is not to provide a good lower bound to the minimum weight, but to help in the location of a reasonable design on the boundary of the permissible region.

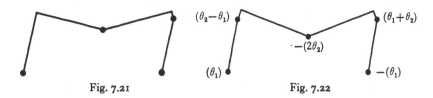

Fig. 7.21   Fig. 7.22

As in the numerical example of section 7.4, the mechanism of fig. 7.21 can now be examined to improve the design. For failure by this mode,

$$4M_A + 2M_B = 1440, \qquad (7.42)$$

and, introducing this into equation (7.39),

$$X = 1440 + 3M_B. \qquad (7.43)$$

Thus M_B should be reduced (from the value of 514) and M_A increased correspondingly. At $M_B = 480$, $M_A = 120$, a hinge is formed at the top of the left-hand column, fig. 7.22, and this is a mechanism of two degrees of freedom which can be tested to see if it satisfies Foulkes's theorem.

From the equation of virtual work, the mechanism of fig. 7.22 taken with the loads of fig. 7.20 gives

$$M_A(2\theta_1 + 2\theta_2) + M_B(2\theta_2) = 240\theta_1 + 1200\theta_2. \qquad (7.44)$$

It will be seen that the weight equation (7.39) can be 'matched' if $\theta_2 = 2 \cdot 5$, $\theta_1 = -0 \cdot 5$, leading to $X = 2880$. However, a negative value for θ_1 is not acceptable at the column feet, and the design corresponding to fig. 7.22, while at a vertex of the permissible region, is not at the minimum-weight vertex.

If the offending hinges are suppressed ($\theta_1 = 0$ in fig. 7.22), the beam mechanism of collapse is left. The work equation gives

$$2M_A + 2M_B = 1200, \tag{7.45}$$

and, using the weight equation (7.39),

$$X = 2400 + M_B. \tag{7.46}$$

Evidently the value of M_B should be further reduced; the next vertex of the permissible region is reached when

$$M_A = M_B = 300, \quad X = 2700. \tag{7.47}$$

The corresponding mechanism, which must be tested to see it it is a proper Foulkes mechanism, is shown in fig. 7.23. The corresponding work equation gives

$$M_A(2\theta_1) + M_B(4\theta_2 - 2\theta_1) = 1200\theta_2, \tag{7.48}$$

and the weight equation (7.39) is matched for $\theta_1 = 2$, $\theta_2 = 2\cdot25$. Since $\theta_2 > \theta_1$, these values are acceptable, and the design of (7·47) is the minimum-weight solution.

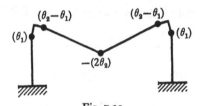

Fig. 7.23

The steps involved in the working of this example may be made quite automatic, and this is discussed in the next chapter. The numerical work can in fact be followed in a sketch of the design plane. There are only six possible mechanisms of collapse, corresponding to the three basic modes and the fact that M_A may be greater or less than M_B. These modes are sketched in fig. 7.24. The corresponding collapse equations are

$$
\left.
\begin{aligned}
(a) \quad & 2M_A + 2M_B = 1200, \\
(b) \quad & 4M_A + 2M_B = 1440, \\
(c) \quad & 4M_A \quad\quad\;\; = 240, \\
(d) \quad & \quad\quad 4M_B = 1200, \\
(e) \quad & 2M_A + 4M_B = 1440, \\
(f) \quad & 2M_A + 2M_B = 240.
\end{aligned}
\right\} \tag{7.49}
$$

Mechanisms (a), (b), (c) and (d) define the boundary of the permissible region, (e) and (f) both lying wholly on the origin side of the region.

The starting-point H is shown in fig. 7.25 as the intersection of mechanism lines (c) and (d). Line OH produced cuts the permissible region at C; the ratio OC to OH is $12/7$. Mechanism (b) was then

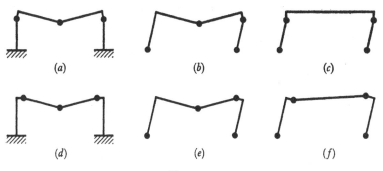

Fig. 7.24

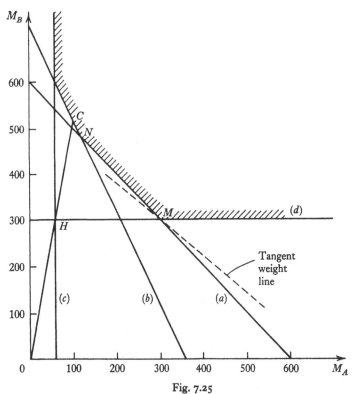

Fig. 7.25

examined, equation (7.42), and a move made to vertex N, fig. 7.22 and equation (7.44). Since this was found not to be the minimum-weight vertex, a move was then made along the line NM, mechanism (a), until the minimum-weight vertex M was reached.

7.6 Alternative loading combinations

The loads shown acting on the portal frame of fig. 7.26 are supposed to represent the working values of dead plus superimposed (V) and of wind load (H). As was noted in chapter 6, it is common practice to design the frame under the action of load V alone to a load factor λ_1, and under the action of both loads V and H to a lower load factor λ_2. Typically the reduction from λ_1 to λ_2 might be 20 or 25 per cent, and, naturally, the more critical of the two cases will govern the design. (*Static* loading only will be considered here; shakedown behaviour is discussed in section 7.7 later.)

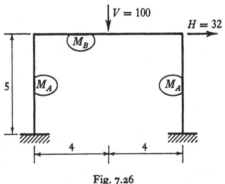

Fig. 7.26

The frame of fig. 7.26 will be designed under the alternative loading combinations $(V, H) \equiv (150, 48)$ and $(200, 0)$, that is, for $\lambda_1 = 2 \cdot 0$ and $\lambda_2 = 1.5$. For the first loading combination, it may be shown by the methods given above that the minimum-weight design is

$$M_A = 120, \quad M_B = 180, \quad X = 1320, \tag{7.50}$$

where, in general,

$$X = 5M_A + 4M_B. \tag{7.51}$$

The corresponding Foulkes mechanism is given by the simultaneous formation of mechanisms (a) and (b) of fig. 7.24. It is easy to see that the design of (7.50) will not carry the other loading combination; the vertical load $V = 200$ requires that the sum of M_A and M_B must be at least 400.

However, the minimum-weight design for the second loading case, considered by itself, is

$$M_A = 0, \quad M_B = 400, \quad X = 1600, \qquad (7.52)$$

and this design is incapable of carrying *any* side load, and hence certainly cannot carry the first loading combination.

Using the mechanisms of fig. 7.24, the permissible regions for the two loading cases are shown in fig. 7.27. The first loading case gives a permissible region bounded by mechanism lines (c), (b), (a) and (d), and it

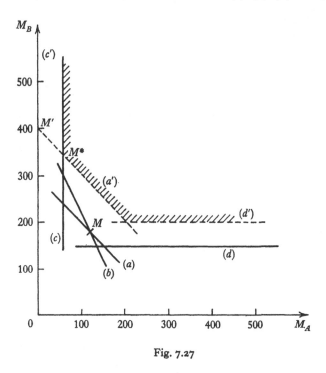

Fig. 7.27

will be seen that the minimum-weight vertex M is *outside* the permissible region (mechanism lines (c'), (a') and (d')) for the second loading case (mechanism (c') is represented by $M_A = 0$). Similarly, the minimum-weight vertex M' for the second region lies outside the first region.

The actual permissible region for the combined loading cases consists of that part of the two individual permissible regions which is common. It is shown hatched in fig. 7.27, and is bounded by mechanism line (c) of the first loading case and mechanism lines (a') and (d') of the second loading case. The minimum-weight vertex of the combined region is at M^*, and the corresponding design is

$$M_A = 60, \quad M_B = 340, \quad X = 1660. \tag{7.53}$$

For this particular numerical example, therefore, Foulkes's mechanism is split between the two loading cases; there *are* two alternative mechanisms, but they are formed one for each loading, and not simultaneously. (It is not necessary, in general, for a collapse mechanism to be formed for each alternative loading pattern that is considered. One or more of the loadings may be so light as to have no influence in determining the minimum-weight design.)

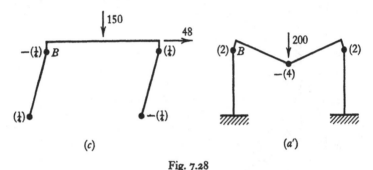

Fig. 7.28

For the present example, Foulkes's mechanism at the vertex M^* of fig. 7.27 is shown in fig. 7.28. The corresponding collapse equations are

$$\begin{aligned} (c) \quad & 4M_A && = 240, \\ (a') \quad & 2M_A + 2M_B && = 800. \end{aligned} \tag{7.54}$$

These two equations can be combined with multipliers α and β to give

$$(4\alpha + 2\beta) M_A + 2\beta M_B = 240\alpha + 800\beta, \tag{7.55}$$

and it is seen that the weight equation (7.51) will be recovered for $\alpha = \frac{1}{4}$, $\beta = 2$, giving $X = 1660$. The corresponding hinge rotations are marked in figs. 7.28. Note that there is no question of any 'cancellation' of the hinge rotation at the knee B of the frame. The absolute values of the hinge rotations must be summed for each span for the individual mechanisms, and then summed again over the two mechanisms (or, in general, over as many mechanisms as are involved). Thus, for the columns, the hinge rotations of fig. 7.28(c) sum to 1, and of fig. 7.28(a') to 4; the total is 5, matching the coefficient of M_A in the weight equation (7.51).

7.7 Shakedown loading

The problem to be discussed involves the ideas of shakedown analysis of chapter 6, and hence the elastic response of the structure is required. Now the elastic bending moments in a frame depend on the relative stiffnesses of the members, and, for the design problem, these stiffnesses are not known at the start of the calculations. Thus, for numerical calculations, there is a vicious circle; the sizes of the members must be known in order to obtain the elastic response, and the elastic response must be known in order to determine the sizes of the members for minimum weight. A complete formal solution could be obtained by working a particular problem in symbols throughout, but such a solution would be unwieldy, even for a computer. An iterative method is suggested at the end of the next chapter, but the general ideas can be illustrated by the simple example of fig. 7.13. This two-span beam will be designed for minimum weight on the assumption that the loads W_A and W_B do not have fixed values, but can vary randomly between the limits

$$
\left. \begin{array}{l} 0 \leqslant W_A \leqslant P_A, \\ 0 \leqslant W_B \leqslant P_B. \end{array} \right\} \tag{7.56}
$$

The linearized weight equation (7.24) will be used as usual. The second moments of area of the two beams will be denoted I_A and I_B, and these quantities will be related in some way to the values of the full plastic moments M_A and M_B. It will be assumed that

$$
\frac{I_A}{I_B} = \left(\frac{M_A}{M_B} \right)^{\gamma}, \tag{7.57}
$$

which implies a geometrical similarity between the two sections. A typical value of the exponent γ might be 1·4, but in the following numerical work γ will be taken as unity for ease of computation; this does not seriously restrict the validity of the results. The elastic solution may be obtained in terms of the stiffnesses I/l of the two spans, and the ratio will be denoted

$$
k = \frac{l_A}{l_B} \cdot \frac{I_B}{I_A}. \tag{7.58}
$$

Table 7.8 gives the maximum and minimum elastic moments at the critical sections 1 to 4 marked in fig. 7.13.

The basic equation (6.28) for incremental collapse may now be applied to table 7.8, using the four possible collapse mechanisms of fig. 7.8. The resulting incremental collapse equations are

$$
\begin{aligned}
(a)\quad 3M_A \qquad\quad &= \frac{P_A l_A}{32}(16) + \frac{P_B l_B}{32}\left(\frac{6}{1+k}\right), \\[2mm]
(b)\quad 2M_A + M_B &= \frac{P_A l_A}{32}(16) + \frac{P_B l_B}{32}\left(\frac{6}{1+k}\right), \\[2mm]
(c)\quad M_A + 2M_B &= \frac{P_A l_A}{32}\left(\frac{6k}{1+k}\right) + \frac{P_B l_B}{32}(16), \\[2mm]
(d)\quad\qquad 3M_B &= \frac{P_A l_A}{32}\left(\frac{6k}{1+k}\right) + \frac{P_B l_B}{32}(16).
\end{aligned}
\tag{7.59}
$$

Table 7.8

Section	$\mathcal{M}^{\max}$	$\mathcal{M}^{\min}$
I	$\dfrac{P_B l_B}{32}\left(\dfrac{3}{1+k}\right)$	$-\dfrac{P_A l_A}{32}\left(\dfrac{8+5k}{1+k}\right)$
2, 3	$\dfrac{P_A l_A}{32}\left(\dfrac{6k}{1+k}\right) + \dfrac{P_B l_B}{32}\left(\dfrac{6}{1+k}\right)$	0
4	$\dfrac{P_A l_A}{32}\left(\dfrac{3k}{1+k}\right)$	$-\dfrac{P_B l_B}{32}\left(\dfrac{5+8k}{1+k}\right)$

For the simple numerical example of fig. 7.6, for which $P_A l_A/32 = 30$ and $P_B l_B/32 = 15$, these equations reduce to

$$
\begin{aligned}
(a)\quad 3M_A \qquad\quad &= 480 + \frac{90}{1+k}, \\[2mm]
(b)\quad 2M_A + M_B &= 480 + \frac{90}{1+k}, \\[2mm]
(c)\quad M_A + 2M_B &= 240 + \frac{180k}{1+k}, \\[2mm]
(d)\quad\qquad 3M_B &= 240 + \frac{180k}{1+k}.
\end{aligned}
\tag{7.60}
$$

Equations (7.60) are directly comparable with (7.20) for the static loading case, and the modifications induced by the variable loading are clearly apparent.

Equations (7.60) are no longer linear in M_A and M_B, since the ratio of stiffnesses k is itself a function of M_A and M_B. For a given expression for

k (such as (7.58) combined with the power law of (7.57)), the equations may be examined by an iterative method. The general form of the equations is not strongly dependent on the expression for k, so that reasonably quick convergence may be expected.

If γ is taken as unity in (7.57), then, for $l_A = l_B$, the value of k is simply M_B/M_A, from equation (7.58). With this linearization, equations (7.60) are plotted in fig. 7.29. The permissible region is now bounded by slightly curved lines, but the general similarity with fig. 7.9 for the static

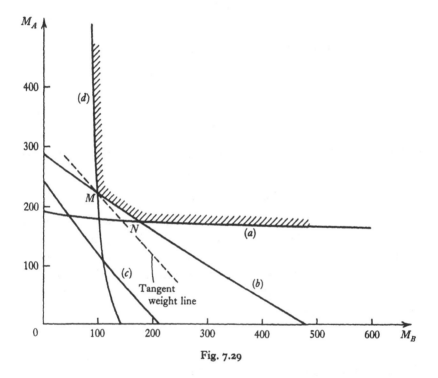

Fig. 7.29

loading case is striking. Still using a linear weight function, the minimum-weight vertex is again located at M, and the minimum-weight design is

$$M_A = 222 \cdot 0, \quad M_B = 98 \cdot 4, \quad X = 320 \cdot 4. \tag{7.61}$$

This represents some 14 per cent increase in weight over the corresponding static case, equations (7.16).

With the curved boundary of the permissible region, there is the possibility of the minimum-weight design no longer occurring at an apex. Thus any computational method which does not actually explore the whole of the boundary of the permissible region, but concentrates on the location of Foulkes mechanisms at the apices, will, in some small number of cases, lead to results which are slightly in error.

Assuming that minimum-weight designs are confined to apices, the possible Foulkes mechanisms for the general two-span beam of fig. 7.13 are those of fig. 7.14. For mechanism (iii), for example, formed when (b) and (d) of equations (7.59) hold simultaneously,

$$
\left.
\begin{aligned}
M_A &= \frac{P_A l_A}{32}\left(\frac{8+7k}{1+k}\right) + \frac{P_B l_B}{96}\left(\frac{1-8k}{1+k}\right), \\
M_B &= \frac{P_A l_A}{32}\left(\frac{2k}{1+k}\right) + \frac{P_B l_B}{96}(16).
\end{aligned}
\right\}
\qquad (7.62)
$$

Now mechanism (iii) occurs only if $M_A \geqslant M_B$, that is, if

$$
\frac{P_A l_A}{P_B l_B} \geqslant \frac{5+8k}{8+5k}.
\qquad (7.63)
$$

At the limit when $M_A = M_B$, then $k = l_A/l_B$ from (7.58), so that (7.63) becomes

$$
\frac{P_A l_A}{P_B l_B} = \frac{5+8(l_A/l_B)}{8+5(l_A/l_B)}.
\qquad (7.64)
$$

Equation (7.64) is plotted in fig. 7.30 as the curve dividing the regions (i) and (iii) from (ii) and (iv). The 'vertical' dividing lines in this figure are more difficult to determine, but implicit analytical expressions may be found; the curves shown in fig. 7.30 have been sketched for the exponent γ in (7.57) equal to unity. Figures 7.15 and 7.30 are again very similar in their general outlines.

7.8 Absolute minimum-weight design

In all the work presented so far in this chapter it has been assumed that members of a frame have been of constant cross-section between joints. The use of prismatic steel members (i.e. standard rolled sections) can be justified for small frames on the count of cheapness of fabrication. For a large frame, however, using made-up rather than standard sections, the designer can vary the cross-section of a member within a span at no great expense. The study of minimum-weight design using members of continuously varying cross-section is therefore of interest. It will be shown that if the members of a frame are designed so that the strength at a particular cross-section corresponds exactly to the bending moment applied at that section, then it is possible to achieve absolute minimum-weight design for the frame.

The establishment of such an absolute minimum gives a fixed point of reference to the designer, and he can estimate penalties in weight and cost should he depart, for practical reasons, from the minimum. It is inconvenient, for example, to vary the cross-section of steel members

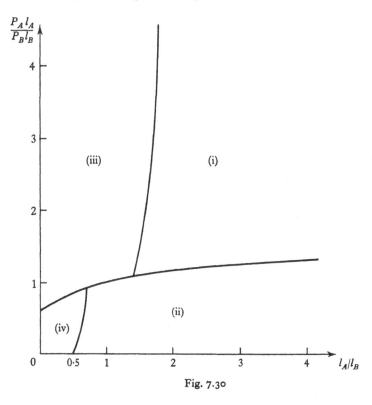

Fig. 7.30

continuously, but abrupt changes in cross-section may imply the use of only a few per cent more material. Reinforced concrete can be 'shaped' more easily than steelwork; here again, the designer can balance the cost of complicated formwork against the saving in material as the minimum-weight design is approached.

For a frame of given dimensions acted upon by a fixed set of given loads, it is possible to construct an infinite number of equilibrium bending-moment distributions, corresponding to the addition to the free bending moments of arbitrary distributions of the reactant moments. The bending moment at a typical section of the frame will be denoted M for one such equilibrium distribution. It is clear that, for this particular distribution, the least weight will occur if the full plastic moment at the typical section is equal to $|M|$; a given distribution of $|M|$ is therefore a possible *design* of the frame.

As before, the weight function will be linearized; to some scale, the total weight of the frame for a particular design is

$$X = \int |M|\, ds, \tag{7.65}$$

where the integration extends over the length of all members. The problem is to find the particular design $|M|$, from all the possible equilibrium distributions, which makes the value of X in equation (7.65) an absolute minimum.

Suppose that a possible deflected shape y for the frame is given by some distribution κ of curvatures; it may be shown (by considering a small variation from the minimum) that the minimum-weight design will deflect in a form containing no discontinuities in slope, so that it is not necessary to consider any localized hinges. The curvatures κ will be said to *conform* with a design $|M|$ if, at each cross-section, the sign of M is the same as the sign of κ; thus a 'hogging' curvature conforms with a 'hogging' bending moment.

Theorem. If, for the given loads, a design $|M^*|$ can be found that conforms with a deflected shape of the structure having *constant* absolute curvature κ_0, then the design $|M^*|$ has minimum weight.

To prove this theorem, suppose that concentrated loads W and distributed loads w act on the frame; by definition, these loads are in equilibrium with the bending moments M^*. Further, suppose that for this design $|M^*|$, the deflexions y of the frame are compatible with curvatures $\pm \kappa_0$, where κ_0 is a constant. Then, by virtual work,

$$\bar{X} \text{ (say)} = \int wy\, ds + \Sigma Wy = \int M^* \kappa_0\, ds. \tag{7.66}$$

Now, since $(\operatorname{sgn} M^*) = (\operatorname{sgn} \kappa_0)$, the quantity $\bar{X}$ is given by

$$\bar{X} = \int |M^*|\, |\kappa_0|\, ds = |\kappa_0| \int |M^*|\, ds = |\kappa_0|\, X^*, \tag{7.67}$$

using the weight equation (7.65). Thus the weight X^* of the putative minimum-weight design is $\bar{X}/|\kappa_0|$.

Now any other design $|M|$, obtained by some other combination of free and reactant moments, is also in equilibrium with the external loads. Thus the quantity $\bar{X}$ of equation (7.66) is also given by

$$\bar{X} = \int M\kappa_0 \, ds. \tag{7.68}$$

Combining equations (7.67) and (7.68)

$$X^* = \frac{\bar{X}}{|\kappa_0|} = \int M \frac{\kappa_0}{|\kappa_0|} \, ds \leqslant \int |M| \left| \frac{\kappa_0}{|\kappa_0|} \right| \, ds, \tag{7.69}$$

the final inequality being necessary since the new distribution M will in general not conform with the curvatures κ_0. Now $\kappa_0/|\kappa_0| = \pm 1$, so that equation (7.69) gives

$$X^* \leqslant \int |M| \, ds = X. \tag{7.70}$$

Thus X^* cannot exceed the weight X of any other design, so that X^* must be the minimum weight.

The theorem may be illustrated by reference to a fixed-ended beam, fig. 7.31. If all the loads act downwards, the free bending-moment diagram will be of the general form of fig. 7.31(b); that is, it will be convex. (Re-entrant bending-moment diagrams lead to slightly more complex calculations, but do not affect the general principles.) The minimum-weight theorem states that the beam must take up a deflected form in which the curvatures are constant. For this simple problem, arcs of constant curvature can be combined only in the way shown in fig. 7.31(c); over the central half of the beam there is an arc of negative curvature, and the two end portions, each of length $\frac{1}{4}l$, have constant positive curvature.

This deflexion diagram is symmetrical, and is the only one which satisfies the theorem; it is completely independent of the applied loads. Now since the signs of the bending moments must be the same as the signs of the curvatures, the inflexion points in fig. 7.31(c) must correspond to changes in sign in the bending-moment diagram. Thus the reactant line is fixed as in fig. 7.31(d). The shaded diagram gives the required values of full plastic moment at each section, and this final minimum-weight design is statically determinate; a central span of length $\frac{1}{2}l$ is suspended from the two end cantilevers each of length $\frac{1}{4}l$. (In practice, the beam section cannot be reduced to zero at the inflexion points; some material must be provided to carry the shear forces, which are neglected in this analysis.)

The deflected form of the minimum-weight design for a propped cantilever is also independent of the loading system; the point of contraflexure must always occur at a distance $l/\sqrt{2}$ from the prop, fig. 7.32. However, this independence holds only for the simplest problems. In general, a knowledge of the loading system is required to locate the

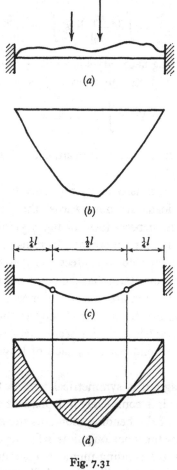

Fig. 7.31

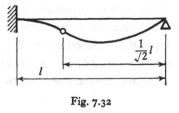

Fig. 7.32

inflexion points. For example, the two-span beam of fig. 7.33 will have two inflexion points, distant (say) $a_1 l_1$ and $a_2 l_2$ from the central support. The values of a_1 and a_2 cannot be determined uniquely from geometrical considerations; general expressions are given below, and it will be found that only one equation can be derived (from the condition of constant curvature) which relates the values of a_1 and a_2. The second required

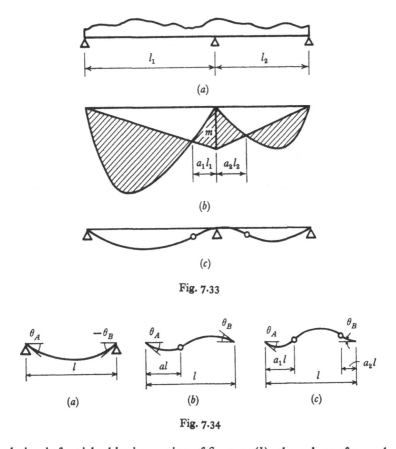

Fig. 7.33

Fig. 7.34

relation is furnished by inspection of fig. 7.33(b); the values of a_1 and a_2 are related by the actual shape of the free bending-moment diagram.

It is always possible to find enough equations to determine the position of the inflexion points in this way, using geometrical relationships arising from the condition of constant curvature, and deducing the remaining equations from the given loading. The geometrical relationships may be most easily written by considering the end slopes of individual spans. In fig. 7.34(a) a single span is bent into an arc of constant curvature with no inflexion point; one inflexion point occurs in fig. 7.34(b), and two in fig. 7.34(c). If the constant curvature κ_0 is taken for convenience as unity, then the following relationships hold:

No inflexion point:
$$\left.\begin{aligned}
\theta_A &= l[\tfrac{1}{2}], \\
\theta_B &= l[-\tfrac{1}{2}].
\end{aligned}\right\} \tag{7.71}$$

One inflexion point:
$$\left.\begin{aligned}
\theta_A &= l[\tfrac{1}{2} - (1-a)^2], \\
\theta_B &= l[\tfrac{1}{2} - a^2].
\end{aligned}\right\} \tag{7.72}$$

Two inflexion points:
$$\left.\begin{aligned}
\theta_A &= l[\tfrac{1}{2} - (1-a_1)^2 + a_2^2], \\
\theta_B &= l[-\tfrac{1}{2} + (1-a_2)^2 - a_1^2].
\end{aligned}\right\} \tag{7.73}$$

These three cases suffice for the solution of most problems. (The fixed-ended beam of fig. 7.31 may be solved by setting $\theta_A = \theta_B = 0$ in equations (7.73), from which $a_1 = a_2 = \tfrac{1}{4}$. Similarly, equations (7.72) lead to the special case of fig. 7.32.) Equations (7.72) applied to the case of fig. 7.33 lead to the single equation

$$(1 - a_1)^2 + (1 - a_2)^2 = 1, \tag{7.74}$$

which arises on matching the slope of the two spans at the internal support. For the numerical example of fig. 7.6, a second relation may be deduced from the general bending-moment diagram of fig. 7.35, namely

$$m = \frac{2a_1}{1 - a_1}(240) = \frac{2a_2}{1 - a_2}(120),$$

or
$$2a_1(1 - a_2) = a_2(1 - a_1). \tag{7.75}$$

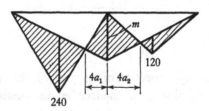

Fig. 7.35

Equations (7.74) and (7.75) solve to give $a_1 = 0.225$, $a_2 = 0.367$, $m = 139$; the bending-moment diagram corresponding to the absolute minimum-weight design of this two-span beam has the proportions sketched in fig. 7.35.

Analytical solutions of more complex problems are really not practicable (although results are quoted below for the rectangular portal frame); numerical methods of solution can be devised, however, which converge rapidly to the correct answer. One such method is an inversion of the standard moment-distribution technique, and is analogous to the elastic method of 'slope distribution'. A start is made by guessing a 'reasonable' reactant line, as, for example, that labelled 'Start' in fig. 7.36 for a long continuous beam of equal spans carrying equal loads. The net bending-moment diagram can then be inspected for any given span, and the positions of the inflexion points read off. The appropriate equations of (7.71) to (7.73) will then give the end slopes of the particular span; it will be found that, in general, these do not match the end slopes of adjacent spans.

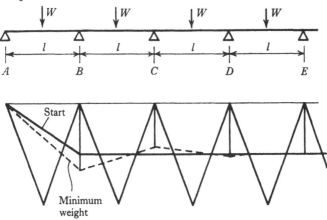

Fig. 7.36

A solution has been found, in fact, which satisfies equilibrium but not compatibility. (The standard moment-distribution technique operates on compatible states which are progressively modified until equilibrium is satisfied.) It will be found, for example, that there is a discontinuity in slope at support B corresponding to the reactant line labelled 'Start' in fig. 7.36; slopes at the other supports match. The discontinuity at B can be eliminated by adjusting the value of the reactant moment at B alone, but this process will introduce a small discontinuity at support C. Adjustment of the reactant moment at C will eliminate this discontinuity, but will introduce a very small discontinuity at B together with one at D. However, only one further 'balancing' will lead to the minimum-weight reactant line sketched in fig. 7.36.

7. MINIMUM-WEIGHT DESIGN

As a final example, the rectangular portal frame of fig. 7.37 will be analysed. There are three different 'modes' for the absolute minimum-weight design of this frame, which occupy, as will be seen, the regions marked in fig. 7.38.

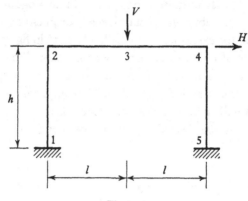

Fig. 7.37

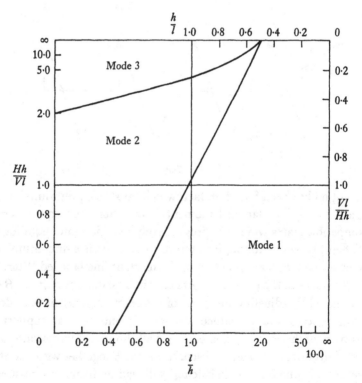

Fig. 7.38

Mode 1. A mode of deformation involving constant curvature of the members of the portal frame is shown in fig. 7.39. The mode is specified in terms of the four points of inflexion by means of the parameters a_1, a_2, b_1 and b_2; the deformations can be specified in terms of rotations θ_2 and θ_4 of joints 2 and 4 and a rotation ϕ measuring the sidesway. Two equations

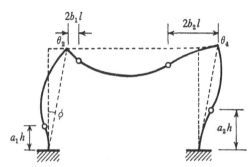

Fig. 7.39

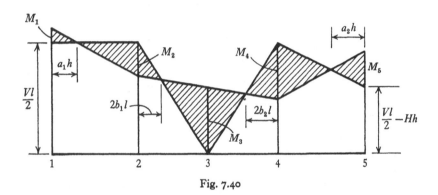

Fig. 7.40

of the type (7.72) or (7.73) can be written for each of the three members, and the rotations θ_2, θ_4 and ϕ eliminated, leaving the three equations

$$\left.\begin{aligned}
1 &= (1-a_1)^2 + (1-a_2)^2, \\
1 - 2a_1 &= \frac{l}{h}[-1 + 2(1-b_1)^2 - 2b_2^2], \\
1 - 2a_2 &= \frac{l}{h}[-1 + 2(1-b_2)^2 - 2b_1^2].
\end{aligned}\right\} \qquad (7.76)$$

The fourth equation necessary for the solution of the problem must be found by considering the bending-moment diagram of fig. 7.40.

The five bending moments M_1 to M_5 are related by two independent equilibrium equations:

$$\left.\begin{aligned} M_2 + 2M_3 + M_4 &= Vl, \\ -M_1 - M_2 + M_4 + M_5 &= Hh. \end{aligned}\right\} \tag{7.77}$$

Further, geometrical consideration of fig. 7.40 gives the relationships

$$\left.\begin{aligned} M_2 &= \frac{2b_1}{1 - 2b_1} M_3, \\[2mm] M_1 &= \frac{a_1}{1 - a_1} M_2, \\[2mm] M_4 &= \frac{2b_2}{1 - b_2} M_3, \\[2mm] M_5 &= \frac{a_2}{1 - a_2} M_4. \end{aligned}\right\} \tag{7.78}$$

Equations (7.77) and (7.78) combine to give

$$\frac{Hh}{Vl}(1 - b_1 - b_2) = \left[\frac{b_2(1 - 2b_1)}{(1 - a_2)} - \frac{b_1(1 - 2b_2)}{(1 - a_1)}\right]. \tag{7.79}$$

Equations (7.76) and (7.79) now suffice to determine the positions of the inflexion points, and hence to fix the bending-moment diagram. The analysis is valid for $b_1 \geqslant 0$; the limit $b_1 = 0$ is shown as the dividing line between modes 1 and 2 in fig. 7.38. For all designs to the right of this line, the values of a_1, a_2, b_1 and b_2 are compatible with the mode sketched in fig. 7.39.

For the particular case $H = 0$, when, by symmetry, $a_1 = a_2$ and $b_1 = b_2$, equation (7.79) is satisfied identically, and equations (7.76) reduce to

$$\left.\begin{aligned} 1 &= 2(1 - a)^2, \\[2mm] 1 - 2a &= \frac{l}{h}(1 - 4b), \end{aligned}\right\} \tag{7.80}$$

from which

$$\left.\begin{aligned} a &= 1 - \frac{1}{\sqrt{2}}, \\[2mm] b &= \frac{1}{4}\left[1 - (\sqrt{2} - 1)\frac{h}{l}\right]. \end{aligned}\right\} \tag{7.81}$$

Now b must be positive, so that mode 1 occurs for $l/h \geqslant (\sqrt{2} - 1)$. For $l/h \leqslant (\sqrt{2} - 1)$, and $H = 0$, mode 2 occurs.

Mode 2. The minimum-weight collapse configuration for mode 2 is shown in fig. 7.41. Comparing with fig. 7.39, the inflexion point near the left-hand knee of the frame has moved from the beam into the column. As before, three equations may be deduced which connect the quantities a_1, a_2, b and c defining the positions of the inflexion points:

$$\left. \begin{aligned} 1+c^2 &= (1-a_1)^2 + (1-a_2)^2, \\ (1-2a_1-2c) &= \frac{l}{h}(1-2b^2), \\ (1-2a_2) &= \frac{l}{h}[2(1-b)^2-1]. \end{aligned} \right\} \tag{7.82}$$

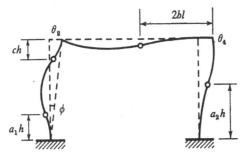

Fig. 7.41

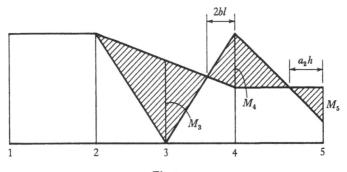

Fig. 7.42

Since two inflexion points occur in the left-hand column, the column section must degenerate to zero; the bending-moment diagram is sketched in fig. 7.42. Consideration of this diagram leads to the fourth equation:

$$\frac{Hh}{Vl}(1-a_2)(1-b) = b. \tag{7.83}$$

The equations are valid providing $c \geqslant 0$, and providing $a_1 \geqslant 0$. The first condition gives the dividing line between modes 1 and 2 in fig. 7.38; the second the dividing line between modes 2 and 3. For the particular case $H = 0$, equation (7.83) gives $b = 0$, and both columns have zero sections.

Mode 3. A mode involving *three* inflexion points is sketched in fig. 7.43; the three equations connecting the values of a, b and c suffice to determine these quantities uniquely. The values of a and c are complementary $(a+c = 1)$, and the inflexion point in the beam is at the centre:

$$a = \frac{1}{2}+\frac{1}{4}\cdot\frac{l}{h}, \quad c = \frac{1}{2}-\frac{1}{4}\cdot\frac{l}{h}, \quad b = \frac{1}{2}. \qquad (7.84)$$

The corresponding bending-moment diagram is sketched in fig. 7.44, and the values of the bending moments may be determined as

$$\left.\begin{aligned}
M_2 &= -\tfrac{1}{2}Vl+\tfrac{1}{2}cHh, \\
M_4 &= \tfrac{1}{2}Vl+\tfrac{1}{2}cHh, \\
M_1 &= \frac{1-c}{c}M_2, \\
M_5 &= \frac{a}{1-a}M_4.
\end{aligned}\right\} \qquad (7.85)$$

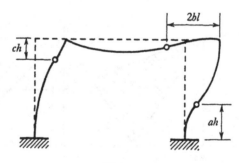

Fig. 7.43

Now the value of M_2 must be positive; this condition gives

$$\frac{Hh}{Vl} \geqslant \frac{1}{c} = \frac{4}{2-(l/h)}. \qquad (7.86)$$

Equality (7.86) is the dividing line between modes 2 and 3 in fig. 7.38.

No other modes occur for this simple frame carrying the idealized loading, so that fig. 7.38 represents the complete range of possible minimum-weight designs. Numerical examples are given as examples 7.15, 7.16 and 7.17 below.

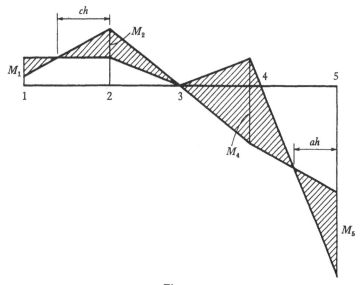

Fig. 7.44

EXAMPLES

Unless stated, a linearized weight function should be assumed in the following examples, together with a continuous range of available sections; spans should be taken to have uniform cross-sections.

7.1 For the continuous beam of fig. 7.3 (*a*), the weight function is

$$X = 4M_A + 4M_B + 6M_C.$$

Determine the values of M_A, M_B and M_C to make the value of X a minimum. (*Ans.* $X = 1920$ for $2M_A + M_B = 480$, $M_B + 3M_C = 480$; $210 \geqslant M_A \geqslant 180$.)

7.2 Design the four-span beam shown for minimum weight. The loads act at the centre of each span. (*Ans.* 200, 200, 450, 100.)

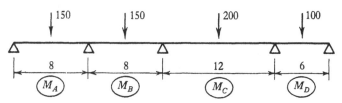

7. MINIMUM-WEIGHT DESIGN

7.3 Design the three-span beam shown for minimum weight for (a) $k = 1$, (b) $k = 2$. The loads act at the centre of each span.

$$(Ans. (a) \tfrac{3}{16}Wl, \tfrac{1}{8}Wl; (b) \tfrac{1}{8}Wl, \tfrac{1}{8}Wl.)$$

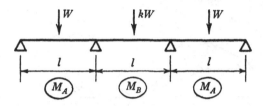

7.4 For the beam shown, confirm that the minimum-weight design is given by

$$M_A = (\tfrac{3}{2} - \sqrt{2})\, wl^2 = 0.686(\tfrac{1}{8}wl^2),$$
$$M_B = (\sqrt{2} - \tfrac{5}{4})\, wl^2 = 1.314(\tfrac{1}{8}wl^2).$$

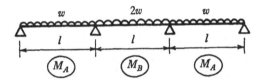

7.5 The long beam shown rests on equally spaced supports; each internal span carries a central point load W, and the end span carries a central point load kW. By considering possible Foulkes mechanisms, show that a uniform beam is not the minimum-weight design except for the special case $k = \tfrac{3}{4}$. For $k \leqslant \tfrac{3}{4}$, show that minimum weight is given by

$$M_A = \tfrac{4}{3}k(\tfrac{1}{8}Wl),$$
$$M_B = (\tfrac{3}{2} - \tfrac{2}{3}k)\,(\tfrac{1}{8}Wl),$$
$$M_C = M_D = \ldots = \tfrac{1}{8}Wl.$$

Similarly, for $k \geqslant \tfrac{3}{4}$, show that

$$M_A = (2k - \tfrac{1}{2})(\tfrac{1}{8}Wl),$$
$$M_B = M_C = \ldots = \tfrac{1}{8}Wl.$$

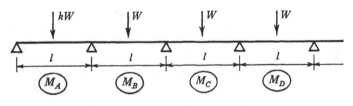

216

7.6 Find the minimum-weight design of the portal frame shown.

(*Ans.* $M_A = 6$, $M_B = 9$.)

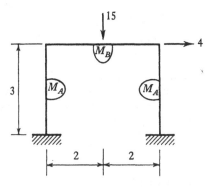

7.7 Find the minimum-weight design of the portal frame shown.

(*Ans.* $M_A = 77$, $M_B = 59$, $M_C = 27$.)

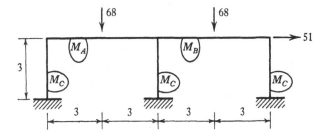

7.8 The plastic moments, $M_R = m_R \lambda PH$ and $M_C = m_C \lambda PH$, of the rafters and columns respectively, in the portal frame shown are to be determined for minimum weight of the structure.

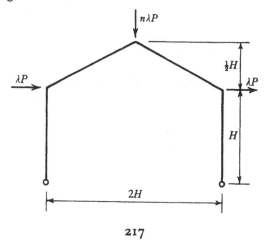

7. MINIMUM-WEIGHT DESIGN

Construct a chart to show the permissible range of designs in the plane of the variables m_R and m_Q, and show that for any value of n a uniform structure gives the most economical design. State the required values of the plastic moments in the cases $0 < n \leqslant 2$ and $n > 2$. (University of London, B.Sc. (Engineering) Part III: Civil, King's College, 1966.) (*Ans. $m = 1$, $m = (n+3)/5$.*)

7.9 Explain why the problem of minimum-weight plastic-collapse design for structures made up of members of uniform sections, having their weights per unit length w related to their plastic moments of resistance M_p by an equation of the form $w = kM_p^n$ (to a very good approximation), can be investigated with little or no loss of accuracy on the assumption that $n = 1$. Also explain how lower and upper bounds for the minimum total weight W_M of such a structure can be established.

The two-bay mild-steel portal frame shown is required to carry the distributed-dead and side-wind loading shown. The beams and columns are to be constant in section with plastic moments of resistance M_1 and M_2 respectively. Show that, if the weight per unit length of a member is taken as kM_p,

$$2 \cdot 5kPl^2 \leqslant W_M \leqslant 3 \cdot 5kPl^2.$$

By means of plastic moment adjustment combined with a mechanism check, or otherwise, determine the values of M_1 and M_2 for minimum weight, assuming any hinge along a beam to occur centrally. (Oxford University, Final Examination in Engineering Science, 1967.) (*Ans. $M_1 = \frac{5}{9}Pl$, $M_2 = \frac{1}{3}Pl$.*)

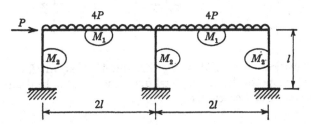

7.10 Determine the minimum-weight design of the two-span beam shown (*a*) on the usual assumption of a linearized weight function: $X = 32M_A + 15M_B$, and (*b*) on the assumption that the weight is given by $X = 32(M_A)^{0.6} + 15(M_B)^{0.6}$. (*Ans. (a) 200, 80; (b) 160, 160.*)

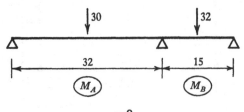

7.11 The frame of fig. 8.2 is to be designed to carry the loading shown, or alternatively, a vertical load only of magnitude 210 units. Determine the values of M_A and M_B for minimum weight. *(Ans. 84, 126.)*

7.12 Show that the minimum-weight design under *static* loading ($V = 24$, $H = 8$) of the frame shown is given by $M_A = M_B = 60$.

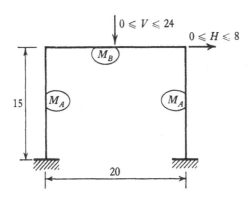

7.13 The loads in example 7.12 vary between zero and the maximum values shown. The flexural rigidity (EI) of a member is related to its full plastic moment (M) by $EI = kM^{1.4}$. Using an iterative method, determine values of M_A and M_B to give minimum weight of the frame against incremental collapse. *(Ans. 56·4, 76·7.)*

7.14 Using the iterative technique suggested for finding the *absolute* minimum weight of the continuous beam shown in fig. 7.36, determine the final position of the reactant line.

$$(Ans.\ M_B = 1\cdot31(\tfrac{1}{8}Wl),\quad M_C = 0\cdot90(\tfrac{1}{8}Wl),\quad M_D = 1\cdot03(\tfrac{1}{8}Wl),$$
$$M_E = 0\cdot99(\tfrac{1}{8}Wl).)$$

7.15 The frame of fig. 7.16 is to be designed (*a*) for minimum weight using prismatic members and with equal columns, $M_A = M_C$, (*b*) for minimum weight using prismatic members but with the columns not necessarily equal, and (*c*) for absolute minimum weight. Show that the bending-moment diagrams at collapse are as shown.

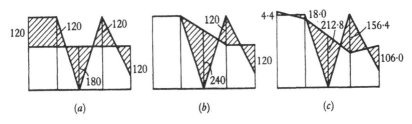

7.16   Repeat example 7.15 with the vertical load reduced to 60 units.

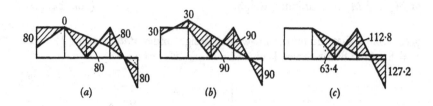

<table>
<tr><td align="center">(a)</td><td align="center">(b)</td><td align="center">(c)</td></tr>
</table>

7.17   Repeat example 7.15 with the vertical load further reduced to 12 units.

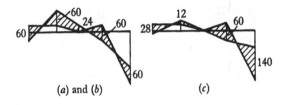

<table>
<tr><td align="center">(a) and (b)</td><td align="center">(c)</td></tr>
</table>

8

NUMERICAL ANALYSIS

This chapter discusses some of the ways in which the ideas of simple plastic theory can be adapted for machine calculation. In most cases programs are already in existence, but the aim of the present discussion is not to give a catalogue of available programs, but to show how the problems of alternative loading systems, for example, or of minimum weight could be set up for numerical solution.

The ideas of dynamic programming were presented in the last chapter in relation to beam systems. Dynamic programming avoids many of the complications of non-linear programming, but the method can in fact operate with strongly non-linear cost functions. It was also mentioned that, since the method is essentially one of search over a given field, any *a priori* reduction of the field will lead to economies in computation time. That is, the more constraints that are introduced into a specific problem the faster the calculations. Toakley† has discussed such problems of 'discrete' programming, including the use of Monte Carlo techniques of random search to find optimum designs.

However, the main discussion here is confined to calculations based on the assumption that a continuous range of sections is available, and will be concerned only with the use of prismatic members between the joints of a frame.

8.1 Linear programming

It was noted in chapter 7 that it was possible to formulate the minimum-weight problem by using the equilibrium and yield conditions only. For the two-span beam of fig. 7.6, the restraining inequalities are

$$\left. \begin{aligned} -M_A &\leqslant 240 - \tfrac{1}{2}m \leqslant M_A, \\ -M_A &\leqslant \qquad m \leqslant M_A, \\ -M_B &\leqslant \qquad m \leqslant M_B, \\ -M_B &\leqslant 120 - \tfrac{1}{2}m \leqslant M_B, \end{aligned} \right\} \tag{8.1}$$

† A. R. Toakley, Optimum design using available sections, *Journal of the Structural Division, Proc. A.S.C.E.*, **94**, no. ST 5, May 1968.

and the weight function to be minimized is

$$X = M_A + M_B. \tag{8.2}$$

All equations and inequalities in (8.1) and (8.2) are linear, and standard computer techniques exist for dealing with such problems in linear programming.

Without discussing these, it is instructive to solve this particular numerical example by hand, using Dines's method.† Suppose a quantity x is confined within bounds:

$$\left.\begin{aligned} a_1 \leqslant x \leqslant b_1, \\ a_2 \leqslant x \leqslant b_2, \\ \cdots \quad \cdots \quad \cdots \\ a_n \leqslant x \leqslant b_n. \end{aligned}\right\} \tag{8.3}$$

Then the necessary and sufficient condition for the existence of x is that all the b's in (8.3) must be greater than all the a's, that is

$$b_p \geqslant a_q \quad \text{(all } p \text{ and } q\text{).} \tag{8.4}$$

A systematic way of writing (8.4) may be deduced from a rewriting of the continued inequalities (8.1) in the form:

$$\left.\begin{aligned} m + 2M_A &\geqslant 480, * \\ m + M_A &\geqslant 0, \\ m + M_B &\geqslant 0, \\ m + 2M_B &\geqslant 240, * \\ \cdots \quad &\cdots \quad \cdots \\ -m + 2M_A &\geqslant -480, \\ -m + M_A &\geqslant 0, \\ -m + M_B &\geqslant 0, * \\ -m + 2M_B &\geqslant -240. \end{aligned}\right\} \tag{8.5}$$

(The significance of the asterisks in these and subsequent inequalities will be explained after (8.11) below.)

Regarding m as the quantity 'x' of (8.3), then the inequalities (8.4) may be generated by adding each of the first four of (8.5) to each of the second

† L. L. Dines, Systems of linear inequalities, *Ann. Math.* **20**, 1918.

four. In this process, m itself is eliminated. The sixteen resultant inequalities are

$$
\left.
\begin{aligned}
4M_A &\geq 0, \\
3M_A &\geq 480, \\
2M_A + M_B &\geq 480, * \\
2M_A + 2M_B &\geq 240, \\
&\cdots \quad \cdots \quad \cdots \\
3M_A &\geq -480, \\
2M_A &\geq 0, \\
M_A + M_B &\geq 0, \\
M_A + 2M_B &\geq -240, \\
&\cdots \quad \cdots \quad \cdots \\
2M_A + M_B &\geq -480, \\
M_A + M_B &\geq 0, \\
2M_B &\geq 0, \\
3M_B &\geq -240, \\
&\cdots \quad \cdots \quad \cdots \\
2M_A + 2M_B &\geq -240, \\
M_A + 2M_B &\geq 240, \\
3M_B &\geq 240, * \\
4M_B &\geq 0.
\end{aligned}
\right\} \tag{8.6}
$$

It will be seen at once that the first of (8.6) can be struck out when compared with the second; similarly, the central eight inequalities are all redundant. In fact, only four out of the sixteen cannot be eliminated at once by cursory inspection; these are

$$
\left.
\begin{aligned}
(a) \quad 3M_A &\geq 480, \\
(b) \quad 2M_A + M_B &\geq 480, * \\
(c) \quad M_A + 2M_B &\geq 240, \\
(d) \quad 3M_B &\geq 240.*
\end{aligned}
\right\} \tag{8.7}
$$

(Closer examination shows that since the first and last of (8.7) give $M_A \geq 160$, $M_B \geq 80$, then the third inequality is also redundant.)

8. NUMERICAL ANALYSIS

Inequalities (8.7) are immediately recognizable as equations (7.20), which represent the four possible modes of collapse, fig. 7.8, and which define the permissible region, fig. 7.9. The minimization of the weight function (8.2) subject to inequalities (8.1) is apparently identical with the problem of the minimization of the same weight function subject to inequalities (8.7). This is in fact a general feature of the linear problem which was pointed out by Chan.† Inequalities (8.1) were established by the use of the equilibrium and the yield conditions, with no reference to the mechanism condition. A single application of Dines's method (there would have been R applications had there been R redundancies) led to the set of inequalities (8.7). This last set could have been established directly (indeed, was so established in chapter 7) by the use of the mechanism and equilibrium conditions, with no reference to the yield condition. This *duality* of approach is, of course, typical of all structural problems; an obvious example lies in the conventional elastic analysis of frames, where 'redundant' quantities can be chosen either as unknown forces or as unknown displacements.

Before continuing the solution of inequalities (8.7) in order to determine the minimum weight of the two-span beam, it may be noted that these inequalities contain the solution to the problem of *analysis* as well as of design. Thus, for a *given* two-span beam, say with $M_A = 2M_p$, $M_B = M_p$, the inequalities become

$$(a) \ M_p \geqslant 80, \quad (b) \ M_p \geqslant 96, \quad (c) \ M_p \geqslant 60, \quad (d) \ M_p \geqslant 80. \quad (8.8)$$

The beam will therefore collapse by mode (b) of fig. 7.8; if the actual M_p provided is, say, 168, then the load factor at collapse would be

$$168/96 = 1 \cdot 75.$$

The method of linear programming can thus form the basis for the collapse analysis of frames; Livesley‡ has described the writing of a program for this purpose. The interest in a specific program for collapse analysis lies mainly in such problems as the systematic choice of redundancies (for a program using the force method), rather than in the exploitation of standard techniques of linear programming.

Returning to the problem of minimum-weight design of the two-span

† H. S. Y. Chan, On Foulkes mechanism in portal frame design for alternative loads, *Journal of Applied Mechanics*, **36**, Series E, no. 1, p. 73, 1969.

‡ R. K. Livesley, The selection of redundant forces in structures, with an application to the collapse analysis of frameworks, *Proc. Roy. Soc.* A, **301**, 493–505, 1967.

beam, the weight function X, equation (8.2), can be introduced into inequalities (8.7) by eliminating M_A (from $M_A = X - M_B$):

$$
\left.
\begin{array}{lll}
(a) & X - M_B \geqslant 160, \\
(b) & 2X - M_B \geqslant 480,^* \\
& \cdots \quad \cdots \quad \cdots \\
(c) & X + M_B \geqslant 240, \\
(d) & M_B \geqslant 80.\,^*
\end{array}
\right\}
\tag{8.9}
$$

Dines's technique can now be used again on the set (8.9) to eliminate M_B:

$$
\left.
\begin{array}{ll}
(a)+(c) & X \geqslant 200, \\
(a)+(d) & X \geqslant 240, \\
(b)+(c) & X \geqslant 240, \\
(b)+(d) & X \geqslant 280.^*
\end{array}
\right\}
\tag{8.10}
$$

Evidently the last of (8.10), marked with an asterisk, is most critical. If this single inequality is satisfied, then this is sufficient to ensure that values of M_A, M_B and m can be found to satisfy the original inequalities (8.1).

Since the value of X is to be made a minimum, the equality sign will be taken in the last of (8.10), that is, $X = 280$. This inequality resulted from the two inequalities starred in set (8.9); these two give, on substituting $X = 280$,

$$
80 \geqslant M_B \geqslant 80.
\tag{8.11}
$$

Thus $M_B = 80$, and hence $M_A = 200$ to give the correct value for X.

The starred inequalities in (8.9) have been picked out in (8.7) and (8.6), and it will be seen that these in turn resulted from the starred members of set (8.5). Inserting the values of M_A and M_B, a unique value of $m\,(=80)$ results. Thus the collapse mechanism of fig. 7.14(iii) has been derived for the minimum-weight design; indeed, Dines's method has located the correct vertex M in fig. 7.9. As a computational technique, however, the method is clearly tedious, even for the simple problem of the two-span beam. The trouble lies in the generation of large numbers of inequalities at each stage in the calculations, many of which are redundant. Thus linear programming techniques for minimum-weight design have attempted to devise rapid methods for locating and testing the vertices of the permissible region.† A more general discussion of the use of non-

† See, for example, R. K. Livesley, The automatic design of structural frames, Quart. Journ. Mech. and Applied Math., vol. 9 part 3, pp. 257–78, 1956.

linear programming techniques for optimal plastic design has been given by Chan.†

A method will now be presented which operates essentially on equilibrium states, and ensures that these satisfy the yield condition, but which also uses the equations for the mechanisms of collapse. The method is automatic, and is designed for machine computation, but some simple examples will be worked here by hand in the hope that the basic ideas will become clear. The calculations are recorded in tables 8.1 to 8.6 below, and it will be found that these tables repay close study. The concern is in fact not so much with the exposition of a numerical technique, but with drawing attention once again to the correspondence between equilibrium statements and virtual mechanisms of collapse. In a sense, the remainder of this chapter attempts to draw together many of the simple techniques used in the plastic analysis and design of plane frames.

8.2 Static collapse analysis

The problem to be discussed is the determination of the collapse load factor of a given frame carrying loads of specified working values (proportional loading). As an example, the portal frame of fig. 8.1 has a uniform full plastic moment of $350\,\mathrm{kN\,m}$, and the collapse load factor λ is to be determined. There are five critical sections in the frame, and

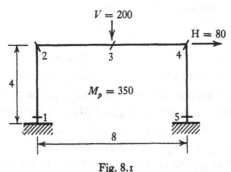

Fig. 8.1

hence there will be two independent equilibrium equations connecting the five values of the bending moments (M_1 to M_5) at these sections. The usual sway and beam mechanisms give

$$\left.\begin{aligned} M_1 - M_2 \qquad + M_4 - M_5 &= 320, \\ M_2 - 2M_3 + M_4 \qquad &= 800. \end{aligned}\right\} \qquad (8.12)$$

† H. S. Y. Chan, Mathematical programming in optimal plastic design, *Int. J. Solids Structures*, vol. 4, pp. 885–95, 1968.

These two equations have been entered in lines 1 and 2 of table 8.1; the coefficients only of the left-hand sides are tabulated in the columns for the appropriate critical sections.

The set of numbers in line 3 of table 8.1, labelled step 'o', represents values of the bending moments at sections 1 to 5 which satisfy equations (8.12). This line is only one of an infinite number of possible equilibrium distributions, and in fact does not represent a very good guess at the bending moments in the frame at collapse. It will be seen that the largest bending moment occurs at section 3, having value 420; thus a frame having $M_p = 420$ will certainly carry the given loads, since line 3 of the table would then satisfy the yield condition. However, the frame would not be on the point of collapse, since only one hinge is formed, and the mechanism condition would not be satisfied. Since the actual full plastic moment of the frame is 350, the collapse load factor cannot be less than $350/420 = 0.83$.

The equilibrium distribution in line 3 will now be modified in such a way that equilibrium is maintained, while the greatest bending moment (420 in this case) is reduced. If changes m_1 and m_5 are made in the values of the moments M_1 to M_5, then only three of these changes can be made arbitrarily; the moments m_1 to m_5 are self-stressing (residual) moments for the frame, and the two independent relationships which must be satisfied may be found from the homogeneous form of equations (8.12):

$$\left. \begin{array}{l} m_1 - m_2 \qquad + m_4 - m_5 = 0, \\ \quad m_2 - 2m_3 + m_4 \qquad = 0. \end{array} \right\} \tag{8.13}$$

Thus, if three values were assigned, say

$$m_3 = m, \quad m_2 = x_1, \quad m_1 = x_2, \tag{8.14}$$

then the values of the other two residual moments, m_4 and m_5, are given by equations (8.13):

$$\left. \begin{array}{l} m_4 = 2m - x_1, \\ m_5 = 2m - 2x_1 + x_2. \end{array} \right\} \tag{8.15}$$

Equations (8.13) are in fact already entered in table 8.1 as lines 1 and 2, if these two lines are now interpreted as homogeneous equations with their right-hand sides zero. Equations (8.14) and (8.15) can thus be derived directly on the table, as shown in line 4; line 4 is displayed as the three lines 5, 6 and 7.

8. NUMERICAL ANALYSIS

Table 8.1

	Section ...	1	2	3	4	5	R.H.S.
1	Step	1	-1	.	1	-1	320
2		.	1	-2	1	.	800
3	'o'	-80	-340	-420	300	240	$\lambda = 0\cdot83$
4		x_2	x_1	m	$2m$ $-x_1$	$2m$ $-2x_1$ $+x_2$	.
5	m	.	.	1	2	2	.
6	x_1	.	1	.	-1	-2	.
7	x_2	1	.	.	.	1	.
8	$m = 40$	.	.	40	80	80	.
9		-80	-340	-380	380*	320	$\lambda = 0\cdot92$
10	m	.	3	1	-1	-4	.
11	x_2	1	.	.	.	1	.
12	$m = 140$	.	420	140	-140	-560	.
13		-80	80	-240	240	-240*	$\lambda = 1\cdot46$
14	m	5	3	1	-1	1	.
15	$m = 40$	200	120	40	-40	40	.
16		120	200*	-200	200	-200	$\lambda = 1\cdot75$

The moments m, x_1 and x_2 are residual moments, and line 4 represents the most general distribution of residual moments for this frame. It will be noted that the moment m 'attacks' the greatest bending moment, 420 at section 3; m will be called the attacking moment, and x_1 and x_2 supplementary moments. Now m, x_1, and x_2 are of course independent; a change can be made in the value of one while keeping changes in the other two zero. The attacking-moment line, line 5 in the table, will be used to modify the equilibrium distribution, line 3, by direct superposition; since the distribution in line 5 is in equilibrium with zero external load, the new line obtained by adding a multiple of line 5 to line 3 will still be an equilibrium distribution for the loaded frame.

If m is chosen to have the value 40, then line 8 shows the residual moments which must be superimposed on line 3 to give the new equilibrium distribution in line 9. The value 40 was chosen because at this value $|M_3| = |M_4| = 380$ in line 9, and this represents the greatest reduction that can be made in the value of the largest of the moments M_1 to M_5. (Note that, as must be the case, the set of numbers in line 9 still satisfies the two original equilibrium equations, lines 1 and 2.)

A frame having $M_p = 380$ would thus sustain the given loads, so that the collapse load factor of the actual frame cannot be less than $350/380 = 0\cdot92$; this is an improvement on the previous lower bound of $0\cdot83$. The frame with $M_p = 380$ and the equilibrium distribution of line 9 has two hinges, at sections 3 and 4, the newly formed hinge being labelled with an asterisk. The rule will be followed that all newly formed hinges will be 'attacked' together with the existing hinges; the bending moment at section 3 continues to be attacked by $+m$, and the moment at section 4 will be attacked by $-m$. Line 4 shows that the residual moment at section 4 is $(2m - x_1)$, so that, following the rule,

$$2m - x_1 = -m,$$

or $$x_1 = 3m. \qquad (8.16)$$

The supplementary moment x_1 is now determined in terms of m, and lines 5, 6 and 7 of the table are replaced by lines 10 and 11 on the elimination of x_1. As before, line 10 is used to attack line 9; taking $m = 140$, line 12, the new equilibrium distribution in line 13 results, in which a third hinge is formed at section 5. (The collapse load factor must thus be at least $350/240 = 1\cdot46$.)

This third hinge is now maintained together with the two previously existing; from lines 10 and 11, the residual moment at section 5 is $(-4m + x_2)$, so that
$$-4m + x_2 = m,$$

or $$x_2 = 5m. \qquad (8.17)$$

The supplementary moment x_2 is thus eliminated, and line 14 gives the new attacking distribution; with $m = 40$, a new hinge is formed at section 2, as shown in line 16, and the corresponding load factor cannot be less than $350/200 = 1\cdot75$.

The residual moment at section 2 is $3m$, from line 14. Following the rule of maintaining the newly formed hinge, $3m$ must be set equal to $-m$, i.e. $m = 0$. Lines 5 to 16 in table 8.1 represent step I of the calculations; step I always ends when $m = 0$, that is, when no further move can be made. It will be seen from line 16 that a beam mechanism has been formed; the hinges at sections 2, 3 and 4 have signs corresponding to the hinge rotations in line 2 of the table. Further, although a hinge is shown at section 5, the full plastic moment of 200 is nowhere exceeded, so that line 16 corresponds to a bending-moment distribution (for a frame having $M_p = 200$) that satisfies the three conditions of mechanism, equilibrium and yield. The collapse distribution has thus been found, and the true load factor is $1\cdot75$.

The calculations in step I always satisfy the conditions of equilibrium and yield, but some orderly step II is required to test whether a mechanism is formed. This step II will be illustrated by means of another numerical example, which also shows that step I does not always proceed in such a straightforward way to give the correct solution.

The frame of fig. 8.1 will be supposed to have a beam section twice the column section, say $M_B = 560$, $M_C = 280$. The calculations will be made as before, except that if a hinge forms at mid-span of the beam, the full plastic moment will be taken as having value $2M_p$, those at all other critical sections having value M_p. Line 17 in table 8.2 represents step '0' for this new example, and is in fact identical with line 3 of table 8.1. A frame with $M_C = \frac{1}{2}M_B = 340$ will therefore certainly carry the loads, the most critical section being at 2. Line 18 displays the most general set of residual moments, with the hinge at section 2 attacked by m. At line 23 a hinge is formed at section 3, and this section must now be attacked by $2m$, so that, from line 18, $x_1 = 2m$. Proceeding as before, a new hinge is formed at section 4 in line 27; from lines 24 and 25, $3m$ must be set equal to $-m$, i.e. $m = 0$, and step I is terminated. It will be noticed that

Table 8.2

#								
17	'0'	.	−80	−340*	−420	300	240	$\lambda = 0\cdot82$
18		.	x_2	m	x_1	$-m +2x_1$	$-2m +2x_1 +x_2$	.
19		m	.	1	.	−1	−2	.
20		x_1	.	.	1	2	2	.
21		x_2	1	.	.	.	1	.
22		$m = 130$	.	130	.	−130	−260	.
23	I	.	−80	−210	−420*	170	−20	$\lambda = 1\cdot33$
24		m	.	1	2	3	2	.
25		x_2	1	.	.	.	1	.
26		$m = 10$	.	10	20	30	20	.
27		.	−80	−200	−400	200*	0	$\lambda = 1\cdot40$
28	II	.	−	−	+	.	.	.
29		Line 2	.	1	−2	1	.	.
30		.	x	$5m$	$2m$	$-m$	$-6m +x$	.
31		m	.	5	2	−1	−6	.
32		x	1	.	.	.	1	.
33	I	$m = 28\cdot6$	.	143	57	−29	−171	.
34		.	−80	−57	−343	171	−171*	$\lambda = 1\cdot63$
35		m	7	5	2	−1	1	.
36		$m = 31\cdot4$	220	157	63	−31	31	.
37			140*	100	−280	140	−140	$\lambda = 2\cdot00$
38	II		+	.	−	+	−	.
39		1+2	1	.	−2	2	−1	.

no further move can be made despite the fact that all the supplementary moments have not been eliminated; this is a common occurrence, and corresponds to the fact that the frame is still redundant.

However, a mechanism of sorts has been developed, as may be seen from step II. In line 28 are noted the signs of the hinges at sections 2, 3 and 4; no hinges are formed at sections 1 and 5. Now lines 1 and 2 of table 8.1 give the hinge rotations of two independent mechanisms, from which can be built up all possible mechanisms of collapse. For the particular stage of the calculations reached in line 28, a mechanism is required having possible hinges at sections 2, 3 and 4, but not at sections 1 and 5. Thus line 1 cannot be used, since the rotations at sections 1 and 5 cannot be 'cancelled' with rotations at the same sections from line 2. However, line 2 itself has no rotations at sections 1 and 5, and the line has been transferred to line 29.

Comparing lines 28 and 29, it will be seen that the mechanism derived by step I has, at section 2, a bending moment of the wrong sign. The rule will be made that, at such a hinge of wrong sign, the restriction that the hinge should be maintained will be dropped. The way is now open for a further operation with step I. (In general, there may be more than one hinge of wrong sign at the end of a step I. Any one such hinge may then be released.)

Line 30 shows section 3 attacked by $2m$ and section 4 by $-m$, thus maintaining these two hinges; the homogeneous equation in line 2 forces the residual moment at section 2 to have the value $5m$, and the other two residual moments can be calculated in terms of a supplementary moment x. Proceeding as before, step I ends at line 37, with hinges formed as shown in line 38. No hinge is formed at section 2; lines 1 and 2 of table 8.1 may be added to give the hinge rotations of line 39, corresponding to the usual combined mechanism of collapse (cf. fig. 1.4(b)). Since the signs of the bending moments and of the hinge rotations now conform, lines 38 and 39, the calculations are complete; the collapse load factor is $280/140 = 2\cdot0$.

Before discussing alternative loading combinations, some features of the above computations may be noted. The *data equations* were entered as lines 1 and 2 in table 8.1; these equations were interpreted as furnishing: (i) the equilibrium equations for the bending moments in the frame acted upon by the given loads, (ii) equilibrium equations for residual moments, and (iii) hinge rotations of elementary mechanisms. In the operation of step I, the attacking moment may always be used to reduce the value of the largest bending moment, and hence to increase the lower bound on the value of the collapse load factor. The original array of residual moments, consisting of an attacking moment m and several supplementary moments x_i, may be considered as redundancies. On the formation of a hinge, one of the supplementary moments may be eliminated; in effect, one of the redundancies is determined since the value of the bending moment is known at the hinge position. Step I always ends when the condition $m = 0$ is reached, but, at this stage (as mentioned above) not all the x_i need have been eliminated, since the frame may well remain partially redundant at collapse.

Step II tests whether a proper mechanism has been formed at the end of step I. It is always possible at the end of step I to combine the original data equations (interpreted as hinge rotations of elementary mechanisms) in such a way that zeros are obtained at the critical sections where no hinges are indicated by step I. (In rare cases more than one combination of mechanisms may be found; this corresponds to overcomplete collapse, in which more than one mechanism of collapse operates simultaneously.) By comparing the mechanism derived from the data equations with the corresponding bending moments at the end of step I, it may be found that the signs of some of the bending moments do not conform with those of the corresponding hinge rotations. Any one of these incompatible hinges is then released from the restriction that the hinge should be maintained; this permits the re-use of step I and, on the first move, the released hinge always disappears. Steps I and II are then used alternately until a step I ends in a bending-moment distribution compatible with a possible mechanism of collapse.

8.3 Static collapse under alternative loading combinations

The vertical and horizontal loads (V and H) in fig. 8.1 will be taken to represent dead plus superimposed and wind loading, respectively. Using load factors of 1·75 and 1·40, the frame might be designed in practice for the worse of the two combinations $(V, H) = (350, 0)$ and $(280, 112)$. The two calculations can be carried on simultaneously, as shown in table 8.3.

Lines 1 and 2 show the data equations, as before. In the five columns headed $\lambda = 1·75$ in row 3 is displayed a bending-moment distribution in equilibrium with $(V, H) = (350, 0)$; the next five columns give a similar distribution for $(280, 112)$. If the beam has a full plastic moment twice that of the columns, the greatest bending moment in line 3, in proportion to the strength of the corresponding critical section, is 476 at section 2. Residual bending-moment distributions are constructed separately for

Table 8.3

Line	Grp	Section	λ = 1·75					λ = 1·4					R.H.S.
			1	2	3	4	5	1	2	3	4	5	
1		Step	1	−1	.	1	−1	.	.	.	.	.	4H
2		Step	.	1	−2	1	.	.	.	.	.	.	4V
3		'o'	0	300	−400	300	0	−112	−476*	−588	420	336	
4			x_5	x_4	x_3	$2x_3 -x_4$	$2x_3 -2x_4 +x_5$	x_2	m	x_1	$-m +2x_1$	$-2m +2x_1 +x_2$	
5		m	.	.	.	.	.	.	1	.	−1	−2	
6		x_1	.	.	.	.	.	.	.	1	2	2	
7		x_2	.	.	.	.	.	1	.	.	.	1	
8		x_3	.	.	1	2	2	.	.	.	.	.	
9		x_4	.	1	.	−1	−2	.	.	.	.	.	
10		x_5	1	.	.	.	1	.	.	.	.	.	
11	I	$m = 176$	.	.	.	.	.	.	176	.	−176	−352	
12			.	0	300*	−400	300*	0	−112	−300	−588	244	−16
13		m	.	−1	−1	−1	.	.	1	.	−1	−2	
14		x_1	.	.	.	.	.	.	.	1	2	2	
15		x_2	.	.	.	.	.	1	.	.	.	1	
16		x_5	1	.	.	.	1	.	.	.	.	.	
17		$m = 6$	.	−6	−6	−6	.	.	6	.	−6	−12	
18			.	0	294	−406	294	0	−112	−294	−588*	238	−28
19		m	.	−1	−1	−1	.	.	1	2	3	2	
20		x_2	.	.	.	.	.	1	.	.	.	1	
21		x_5	1	.	.	.	1	.	.	.	.	.	
22		$m = 14$	.	−14	−14	−14	.	.	14	28	42	28	
23			.	0	280	−420	280	0	−112	−280	−560	280*	0
24	II		.	+	.	+	.	−	−	.	+	.	
25		2	.	.	.	.	.	.	1	−2	1	.	
26	I		x_5	$-m$	$-m$	$-m$	x_5	x_2	$5m$	$2m$	$-m$	$-6m +x_2$	
27		m	.	−1	−1	−1	.	.	5	2	−1	−6	
28		x_2	.	.	.	.	.	1	.	.	.	1	
29		x_5	1	.	.	.	1	.	.	.	.	.	
30		$m = 40$	.	−40	−40	−40	.	.	200	80	−40	−240	
31			.	0	240	−460	240	0	−112	−80	−480	240	−240*
32		m	.	−1	−1	−1	.	7	5	2	−1	1	
33		x_5	1	.	.	.	1	.	.	.	.	.	
34		$m = 6·7$	.	−7	−7	−7	.	47	33	13	−7	7	
35			.	0	233	−467*	233	0	−65	−47	−467	233	−233

the two loading cases as shown in line 4, and these are displayed as lines 5 to 10.

Proceeding as before, line 11 shows the changes resulting from $m = 176$, and line 12 gives the new equilibrium bending-moment distributions.

It will be seen that two hinges are formed at sections 2 and 4 for the loading case $\lambda = 1\cdot75$. At this stage, it is certain that a frame with columns of full plastic moment 300 and beam of full plastic moment 600 will carry either load combination. As before, newly formed hinges will be maintained; this condition requires

$$\left. \begin{aligned} x_4 &= -m, \\ 2x_3 - x_4 &= -m, \end{aligned} \right\} \tag{8.18}$$

and the modified residual moments are displayed in lines 13 to 16. The moment m now attacks both loading combinations.

Step I ends at line 23. Application of step II shows that the hinge should be released at section 2 for the wind-load combination; step I is then reapplied and ends in the beam mechanism for $\lambda = 1\cdot75$ in line 35. A frame having columns with full plastic moment 233 and a beam of full plastic moment 467 will just carry the vertical load at a load factor of $1\cdot75$; moreover, an equilibrium bending-moment distribution satisfying the yield condition has been derived in the last five columns of line 35 for the wind loading at a load factor of $1\cdot40$. Thus the frame will certainly carry this second load combination.

The form of computation shown in table 8.3 will pick out automatically the most critical load combination acting on a frame; if more than two load combinations are specified, then obviously they can all be considered simultaneously.

8.4 Shakedown analysis

It will be remembered that necessary and sufficient conditions for a frame to shake down under random variable loading could be expressed by means of inequalities of the form

$$\left. \begin{aligned} \mathcal{M}_i^{\max} + m_i &\leqslant (M_p)_i, \\ \mathcal{M}_i^{\min} + m_i &\geqslant -(M_p)_i, \end{aligned} \right\} \tag{8.19}$$

together with a third inequality (which will not be considered further here) to guard against alternating plasticity. If any set of residual moments m_i can be found for the frame which satisfies (8.19) at every critical section, then the frame will shake down. Now the tabular method of computation presented in the preceding sections operates by adjusting the residual moments, and so can be adapted easily for shakedown analysis. The step I will consist essentially in modifying inequalities (8.19), and step II in testing whether a mechanism of incremental collapse has been formed.

As an example, the uniform rectangular portal frame of fig. 6.9 will be analysed again, with the loads V and H varying between the limits

$$\left.\begin{array}{c} 0 \leqslant V \leqslant 90, \\ 0 \leqslant H \leqslant 60, \end{array}\right\} \qquad (8.20)$$

and with $M_p = 180$; values of loads and full plastic moment have been multiplied by 3 to avoid fractions in the table of computations.

The maximum and minimum values of the elastic bending moments at the five critical sections are given in table 6.3; these values have been entered ($\times 3$) as line 3 of table 8.4. The largest absolute value in line 3 is 117, so that a uniform frame having $M_p = 117$ will certainly eventually resist all variations of the applied loads purely elastically. Since the actual full plastic moment provided is 180, the shakedown load factor λ_s is given by $\lambda_s \geqslant 180/117 = 1 \cdot 54$.

Exactly the same technique as before will be used in order to improve the lower bound on λ_s. Line 4 of the table shows the most general distribution of residual moments, in which m attacks the moment 117. The residual distributions are the same in both halves of table 8.4, since *one* such distribution must be found to satisfy (8.19); that is, the m_i is the same in both inequalities (8.19). The work proceeds as before, except that hinges can be formed in either half of the table. Step I ends at line 16. Line 17 shows the signs of the bending moments, and these are combined in line 18; line 19 shows that these signs are compatible with the combined mechanism of collapse. Thus this mode of incremental collapse will just occur at $\lambda_s = 180/106 = 1 \cdot 70$; cf. equations (6.42) and fig. 6.10. Had lines 18 and 19 in table 8.4 been incompatible, then the release of a hinge of the wrong sign would have permitted a further step I.

As an example of failure by alternating plasticity, the same frame (fig. 6.9, $M_p = 180$) will be analysed under the loads

$$\left.\begin{array}{c} 0 \leqslant V \leqslant 75, \\ -80 \leqslant H \leqslant 80. \end{array}\right\} \qquad (8.21)$$

Table 8.5 gives the working; from the last line, it will be seen that alternating hinges are formed at the column feet if the load factor exceeds $180/115 = 1 \cdot 57$. If the load factor *did* exceed this value, then it would not be possible to find a set of residual moments for the frame which remained fixed as the applied loads varied. In fact, the range of moment at the column feet, 230 from line 3 of the table, controls the behaviour of the

Table 8.4

Section ...		Maximum					Minimum					
		1	2	3	4	5	1	2	3	4	5	
1	Step	1	−1	.	1	−1	.	.	.	.	.	
2	Step	.	1	−2	1	.	.	.	.	.	.	
3	'o'	75	72	0	117*	0	−36	−45	−108	0	−111	$\lambda = 1\cdot54$
4		x_2	$m+2x_1$	x_1	$-m$	$-2m-2x_1+x_2$	x_2	$m+2x_1$	x_1	$-m$	$-2m-2x_1+x_2$	
5	m	.	1	.	−1	−2	.	1	.	−1	−2	
6	x_1	.	2	1	.	−2	.	2	1	.	−2	
7	x_2	1	.	.	.	1	1	.	.	.	1	
8	$m=2$	.	2	.	−2	−4	.	2	.	−2	−4	
9	I	75	74	0	115	−4	−36	−43	−108	−2	−115*	$\lambda = 1\cdot57$
10	m	3	1	.	−1	1	3	1	.	−1	1	
11	x_1	2	2	1	.	.	2	2	1	.	.	
12	$m=7$	21	7	.	−7	7	21	7	.	−7	7	
13		96	81	0	108	3	−15	−36	−108*	−9	−108	$\lambda = 1\cdot67$
14	m	5	3	1	−1	1	5	3	1	−1	1	
15	$m=2$	10	6	2	−2	2	10	6	2	−2	2	
16		106*	87	2	106	5	−5	−30	−106	−11	−106	$\lambda = 1\cdot70$
17	II	.	+	.	.	+	.	.	.	−	.	
18	II	.	+	.	−	+	−	.	.	.	.	
19	1+2	1	.	−2	2	−1	.	.	.	.	.	

Table 8.5

Section ...		Maximum					Minimum				
		1	2	3	4	5	1	2	3	4	5
1	Step	1	−1	.	1	−1	.	.	.	.	.
2	Step	.	1	−2	1	.	.	.	.	.	.
3	'o'	100	120	0	120	100	−130*	−60	−90	−60	−130
4		m	x	x	x	m	m	x	x	x	m
5	m	1	.	.	.	1	1	.	.	.	1
6	x	.	1	1	1	.	.	1	1	1	.
7	$m=10$	10	.	.	.	10	10	.	.	.	10
8	I	110	120*	0	120*	110	−120	−60	−90	−60	−120
9	m	1	−1	−1	−1	1	1	−1	−1	−1	1
10	$m=5$	5	−5	−5	−5	5	5	−5	−5	−5	5
11		115*	115	−5	115	115*	−115	−65	−95	−65	−115
12	II	+	+	.	+	+	−	.	.	.	.
13	II	±	+	.	+	±	.	.	.	.	−

frame rather than incremental collapse. If this total range is to be accommodated within the value $2M_y$, rather than $2M_p$, then the permissible load factor would be less than $1\cdot57$.

8.5 Minimum-weight design

The tabular method presented in the previous sections is particularly powerful for automatic computation of the minimum-weight problem, that is, as an alternative (which, however, has some of the features of the Dual Simplex algorism) to linear programming. As usual, a numerical example worked by hand will illustrate the techniques used.

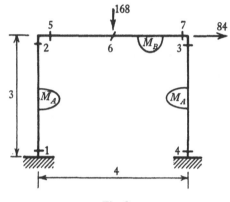

Fig. 8.2

The frame of fig. 8.2 is to be designed to carry the given loading, with both columns having the same section, full plastic moment M_A, but with possibly a different section for the beam, full plastic moment M_B. Thus the linearized weight function to be minimized is

$$X = 6M_A + 4M_B. \tag{8.22}$$

Seven critical sections are marked in fig. 8.2; the duplication at the knees of the frame is necessary to allow for $M_A \gtrless M_B$. There must therefore be four independent equilibrium equations connecting the values of the bending moments at the critical sections; that is, there must be four independent mechanisms of collapse. These will be taken as the usual sway and beam equations, together with two elementary joint rotations; these four mechanisms are entered (in coefficient form) as lines 1 to 4 in table 8.6. (The right-hand sides of the equations correspond to the loading in fig. 8.2.) Step 'o' in table 8.6 gives a (somewhat arbitrary) set of numbers which satisfies the four basic data equations, and hence corresponds to a possible equilibrium bending-moment distribution for the frame.

The critical sections at which the full plastic moment is the same have been grouped together in table 8.6, so that it can be seen at once that a design of frame having $M_A = 252$, $M_B = 188$, $X = 2264$ will certainly

Table 8.6

Full plastic moment ...		M_A				M_B			R.H.S.
Section ...		1	2	3	4	5	6	7	R.H.S.
1	.	1	−1	1	−1	.	.	.	252
2	Step	.	.	.	.	1	−2	1	336
3		.	.	1	.	.	−1	.	.	0
4		.	.	.	1	.	.	−1	0
5	'o'	252*	−30	−10	20	−30	−188*	−10	X = 2264
6	6 × 136 = 816, m_A	−1	.	.	−1	.	.	.	.
7	4 × 66 = 264, m_B	.	.	2	2	.	1	2	.
8	x	.	.	1	−1	−2	1	.	−1	.
9	$m_A = 136$	−136	.	.	−136	.	.	.	.
10	.	116	−30	−10	−116*	−30	−188	−10	X = 1448
11	6 × 43 = 258, m_A	−1	−1	1	1	−1	.	1	.
12	I 4 × 99 = 396, m_B	.	1	1	.	1	1	1	.
13	$m_B = 99$	.	99	99	.	99	99	99	.
14	.	116	69	89	−116	69	−89	89*	X = 1052
15	8 × 9 = 72, m_B	2	3	−1	−2	3	1	−1	.
16	$m_B = −9$	−18	−27	9	18	−27	−9	9	.
17	.	98	42	98*	−98	42	−98	98	X = 980
18	.	+	.	+	−	.	−	+	.
19	α, 1+2+3	1	.	1	−1	.	−2	1	.
20	II β, 4	.	.	1	.	.	.	−1	.
21	.	(α)	.	(α+β)	−(α)	.	−(2α)	(α−β)	.
22	α = 5/3, β = 1	(5/3)	.	(8/3)	−(5/3)	.	−(10/3)	(2/3)	.
23	$m_A = 43$	−43	−43	43	43	−43	.	43	.
24	.	73	−73*	33	−73	−73	−188	33	X = 1190
25	I 14 × 10 = 140, m_A	−1	1	3	1	1	2	3	.
26	$m_A = 10$	−10	10	30	10	10	20	30	.
27	.	63	−63	63*	−63	−63	−168	63	X = 1050
28	.	+	−	+	−	.	−	.	.
29	α, 1	1	−1	1	−1	.	.	.	.
30	II β, 2+3+4	.	1	1	.	.	−2	.	.
31	.	(α)	−(α−β)	(α+β)	−(α)	.	−(2β)	.	.
32	α = 3/2, β = 2	(3/2)	−(−1/2)	(7/2)	−(3/2)	.	−(4)	.	.
33	2 × 35 = 70, m_A	−1	−3	−1	1	−3	−2	−1	.
34	I $m_A = −35$	35	105	35	−35	105	70	35	.
35	.	98	42	98	−98	42	−98	98*	X = 980

carry the loads. In order to modify this design, the equilibrium distribution will be 'attacked' by a general distribution of residual moments, exactly as before. However, instead of a single attacking moment m, as many attacking moments will be used as there are 'spans' to design. In this example, the residual moment at section 1 will be denoted by $-m_A$, and at section 6 by m_B; since the frame has three redundancies, one supplementary moment x is required, and this has been placed (arbitrarily) at section 2.

Thus the homogeneous forms of the basic equations in lines 1 to 4 of the table 8.6 are

$$\left. \begin{aligned} m_1 - m_2 + m_3 - m_4 &= 0, \\ m_5 - 2m_6 + m_7 &= 0, \\ m_2 - m_5 &= 0, \\ m_3 - m_7 &= 0, \end{aligned} \right\} \tag{8.23}$$

where m_1, m_2 and m_6 have been assigned as

$$m_1 = -m_A, \quad m_2 = x, \quad m_6 = m_B. \tag{8.24}$$

The other residual moments (m_3, etc.) can now be found in terms of the values (8.24), and the general distribution of residual moments is entered as lines 6, 7 and 8 in table 8.6.

Either m_A or m_B can be used in table 8.6 to reduce the value X of the weight. A unit change in m_A produces a change of 6 in X, and the largest change that can be made is $m_A = 136$, implying a weight reduction of 816 units; at this value of m_A, line 9, the equilibrium distribution is modified so that a new hinge is formed at section 4, as shown in line 10. Had m_B been used to reduce the value of M_B, the largest reduction in X would have been $4 \times 66 = 264$, so that a greater reduction results from attacking span A first.

As usual, the newly formed hinge in line 10 will be attacked by m_A; the degree of redundancy in the problem is reduced by 1, and the supplementary moment x can be eliminated to leave the pattern of residual moments in lines 11 and 12. It is now more efficient to attack span B, as shown in lines 13 and 14. The newly formed hinge in line 14 requires that $(m_A + m_B)$ be set equal to $-m_B$, that is $m_A = -2m_B$, and the single pattern now possible is shown in line 15.

Weight is now reduced if M_B is increased; the design of the two spans is not now proceeding independently. An increase of one unit in M_B is accompanied by a decrease of 2 units in M_A, as may be seen from line 15; the value of X is correspondingly reduced by 8 units, from equation (8.22).

Step I ends at line 17, no further moves being possible. A vertex of the permissible region has been reached, and this vertex must be tested to see whether it corresponds to a Foulkes mechanism.

The hinge rotations in lines 1 to 4 must be combined in such a way that no hinges appear at critical sections 2 and 5; the addition of lines 1, 2 and 3 achieves this, and line 4 involves hinges at critical sections 3 and 7 only. Thus lines 19 and 20 of the table indicate that there are two simultaneous collapse mechanisms formed at the vertex; these will be superimposed with multipliers α and β, as shown in line 21. Now if the mechanism is a Foulkes mechanism, the quantities in brackets in each span must sum numerically to the length of that span; that is, using the coefficients of equation (8.22),

$$\left.\begin{aligned} 3\alpha + \beta &= 6, \\ 3\alpha - \beta &= 4, \end{aligned}\right\} \tag{8.25}$$

from which $\alpha = \frac{5}{3}$, $\beta = 1$, as noted in line 22 of the table. All quantities inside the brackets are positive, so that all hinges are of the correct sign, and the minimum weight vertex has been found.

In general, step I will end in a mechanism of the required number of degrees of freedom, but which is not a Foulkes mechanism; that is, the wrong vertex will have been reached. If line 10 in table 8.6 had been modified by attacking M_A rather than M_B, then the calculations would have followed the course shown in lines 23 to 27. Step I ends with five hinges, line 28, and the two mechanisms are shown in lines 29 and 30. The required hinge rotations for Foulkes's mechanism are shown in line 32, but the sign of the rotation at critical section 2 is not compatible with the sign of the bending moment there. This hinge is therefore 'released', thus permitting the design to be modified; the new step I is shown in lines 33, 34 and 35, and ends at the minimum-weight vertex.

Just as for the case of analysis, it can be shown that steps I and II in minimum-weight design can always be carried on in the way illustrated by table 8.6, and that the weight X of a frame can thereby be progressively reduced. It can happen by accident that *two* hinges are formed simultaneously on the last move of a step I; if the correct vertex has not been reached, then *two* hinges (whose positions are indicated by the analysis) must be suppressed in order that a new step I can be initiated.

The techniques outlined here have been programmed successfully†;

† J. J. Kalker, Automatic minimum weight design of steel frames on the IBM 704 Computer, *Rept. IBM* 2038/3, Brown University, Providence RI, 1958.

it is clear that the same kind of calculations can be made for the case of minimum-weight design under (static) alternative loading combinations. Similarly, *design* under shakedown conditions can also be made automatic, although, as indicated in section 7.7 earlier, an iterative method must be used. The calculation of tall buildings is a specialized topic, to which brief reference is made at the end of the next chapter.

9

MULTI-STOREY FRAMES

The type of frame to be discussed in this chapter is one which is tall enough for wind loading to be an important factor in design, and for which stability considerations may be critical. Thus a single-bay single-story frame of normal dimensions will not be tall in this sense, although the unusual frame of three bays and two storeys for which calculations are made in the next chapter is seriously affected by wind. A fifty-storey ten-bay office block is certainly tall, and so also is a ten-storey two-bay frame, but these two frames may respond to their loadings in different ways.

It is usual in the design of such tall buildings to attempt to carry the wind load to the foundations by means of bracing in the planes of the frames. Such designs are discussed in sections 9.2 and 9.4 below. However, it may prove impossible (perhaps for architectural reasons) to incorporate bracing in the structure; in the design example of chapter 10, bracing cannot be used because electrical clearances would be insufficient. All loads on the structure must then be resisted by bending of the beams and columns, and this type of frame is discussed in section 9.5. In both types of design, braced and unbraced, there are certain common features which may be discussed conveniently as a preliminary to detailed calculations.

9.1 Design considerations

As was stated in chapter 6 of vol. 1, it is the task of the designer to construct a 'reasonable' equilibrium state for his structure on which to base a design. An elastic distribution of bending moments may be quite acceptable for this purpose, but to obtain such a distribution may not be easy. It is true that the use of computers makes the elastic analysis of a given frame under given loads relatively straightforward, but any particular building may be subject to a large number of different loading combinations. Solutions must be found for each of these, and they must then be superimposed to give the most critical values for design.

By contrast, it is easy to show that, for frames acted upon by gravity loading and horizontal loading in one direction, then *all* loads should act together to give the worst case for *simple plastic* design. There is no question of some beams being loaded and others unloaded; the full dead plus superimposed loads should be applied everywhere. (For elastic column design, as will be seen below, alternative loading patterns must be considered. For the primary plastic design, however, the complications of using anything analogous to influence coefficients are avoided.)

The use of plastic methods has another and much more important advantage. Elastic theory is concerned essentially with *analysis*; it is difficult, even with a computer, to introduce member sizes as unknown variables. That is, an elastic solution requires a prior knowledge of the elastic moment–curvature response of each member. At collapse, however, as has been noted throughout these two volumes, the portion of the frame concerned becomes *statically determinate*; bending moments can be found without the use of the moment–curvature relation and without using any statements about the compatibility of the deformations (beyond the geometrical requirements of a mechanism of collapse). This makes it possible to devise methods of direct design rather than analysis, and to go a long way to the elimination of iterative processes. In so far as elastic checks may have to be made, it is possible that a plastic design may have to be successively modified; the purely plastic work is, however, direct.

The application of simple plastic theory to the design of tall buildings exposes the limitations that result from the basic assumptions of the theory. For example, deflexions are assumed small, so that the equilibrium equations are not affected by the actual behaviour of the building. Deflexions of a braced frame *will* be small, but a tall unbraced frame could develop, at least in theory, sufficiently large lateral displacements that extra bending moments are induced in the columns. This point will be discussed in a moment in conjunction with one of the other basic assumptions, that of stability.

The introduction of unstable elements into a steel frame is really inadmissible by any design theory. The fact that columns are potentially unstable is something that must be considered very seriously by the designer; he must ensure that the structure as a whole finally becomes unserviceable in a stable way, for example by plastic collapse of beams rather than catastrophic collapse of columns. In fact, the condition that columns in a tall building should be designed to remain stable turns out to be not very demanding. The axial loads in all but the upper few column lengths are high, and this ensures that a fairly heavy section must be used having small slenderness ratios about either axis. (Exceptionally long columns, for example in a ground-floor entrance hall, will obviously be more critical.) If the frame is unbraced, then the large bending moments induced by wind will require even heavier column sections, so that again local instability is no real problem.

In addition, the frame as a whole must not become unstable. The phenomenon of *frame instability* may perhaps be introduced by reference to the example of a simple rectangular portal frame, fig. 9.1. The frame is of uniform section, full plastic moment M_p and flexural rigidity EI, and carries vertical and horizontal loads W at the usual idealized loading points. Collapse according to the simple (rigid–plastic) theory will occur by the combined mode with hinges at A, C, D and E; the corresponding load at collapse is $W = 3M_p/l$.

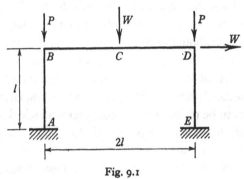

Fig. 9.1

The frame also carries two concentrated loads P at the top of each column; for simplicity, the value of P is supposed to be so large compared with W that the axial load in each column can be taken as P. Ignoring for the moment any effects due to these axial loads, an *elastic*–plastic analysis may be made by the methods of chapter 5 to determine the order of formation of the plastic hinges as the value of W is increased slowly from zero. It will be found that the following results are obtained:

$$\left. \begin{aligned} \text{Hinge at } E: \quad W &= \frac{80}{33}\frac{M_p}{l} = 2\cdot42\frac{M_p}{l}. \\[2mm] \text{Hinge at } D: \quad W &= \frac{172}{67}\frac{M_p}{l} = 2\cdot57\frac{M_p}{l}. \\[2mm] \text{Hinge at } C: \quad W &= \frac{68}{23}\frac{M_p}{l} = 2\cdot96\frac{M_p}{l}. \\[2mm] \text{Hinge at } A: \quad W &= 3\frac{M_p}{l}. \end{aligned} \right\} \qquad (9.1)$$

This analysis assumes, as usual, that there is no spread of plastic zones at the hinges, and that strain hardening can be neglected.

When the value of W is zero, there will be a certain critical value of the loads P in fig. 9.1 which will cause the frame as a whole to become unstable in a sway mode, fig. 9.2 (a). (It is assumed that instability out of the plane of the frame is prevented.) The critical value of P can be calculated in any of the standard ways (for example, by the use of s and c functions), and will be found to be $0.612P_e$, where $P_e(= \pi^2 EI/l^2)$ is the buckling load for a column of the same properties but with pinned ends. After the formation of the first plastic hinge, at the right-hand column foot E, the elastic

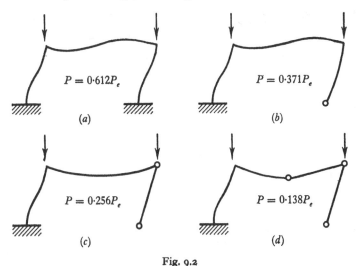

$$P = 0.612P_e$$

$$(a)$$

$$P = 0.371P_e$$

$$(b)$$

$$P = 0.256P_e$$

$$(c)$$

$$P = 0.138P_e$$

$$(d)$$

Fig. 9.2

restraint at this footing will no longer be available to resist the development of any unstable sway mode, and the critical value of P drops from $0.612P_e$ to $0.371P_e$ for the frame of fig. 9.2 (b). The formation of a second plastic hinge at D turns the frame (for the purpose of stability analysis) into that shown in fig. 9.2 (c), for which the critical load is $0.256P_e$. This *deterioration of stability* continues with the formation of the third hinge, fig. 9.7 (d). Finally, the collapse condition $W = 3M_p/l$ corresponds to a mechanism for which the elastic critical load P is zero.

On this simple analysis, therefore, it is to be expected that the formation of the last hinge should be accompanied by a run-away type of collapse. The fact that this does not necessarily happen in practice is due to the effect of strain hardening, which can stiffen the collapsing frame; moreover, axial loads may not be high for low buildings. In multi-storey frames, however, axial loads may be large; the loads P in fig. 9.1 might represent the presence of a number of storeys above the first. Further, and acting against the effect of strain hardening, geometry changes (due to sway) may become significant; once the storey sways cannot be ignored, the axial loads in the columns will contribute to the bending moments in the columns (the so-called '$P\Delta$' effect).

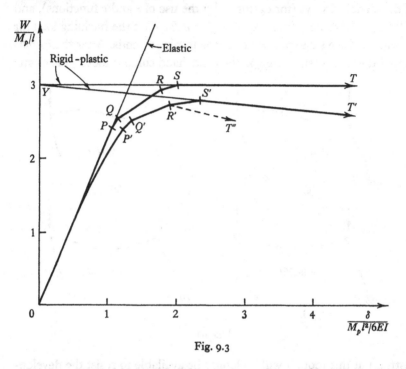

Fig. 9.3

These various effects are displayed in fig. 9.3, which shows the behaviour of the frame of fig. 9.1 in the presence of the axial loads P as the loads W are increased slowly until collapse occurs. According to the simple rigid–plastic analysis, no deformation at all will occur until the collapse load, $3M_p/l$, is reached, and the load–deflexion curve will consist of the two straight lines OY and YT. The straight lines $OPQRST$ record the results of the simple elastic–plastic analysis, equations (9.1), ignoring all changes of geometry and instability effects; the (small) sidesway deflexion δ is plotted against the value of W.

The '$P\Delta$' effect may now be introduced. On the rigid–plastic theory, the geometry changes could well imply a fall in the value of W as the sidesway deflexion increases, and the load–deflexion curve would then

246

be modified from OYT to OYT'. If now a full elastic–plastic analysis is made, allowing for geometry changes, the predicted load–deflexion curve might be that shown as $OP'Q'R'S'T'$, the points P', Q', R' and S' representing (as before, although the order might conceivably be changed) the formation of hinges E, D, C and A.

Still neglecting strain hardening, the behaviour at S' is clearly unstable; instead of a quasi-static collapse process, the formation of the last hinge at S' would imply a run-away condition. So far, however, the ideas recorded in fig. 9.2 have not been introduced. It might be that on the formation of the third hinge at R' in fig. 9.3 (and equally at R if geometry changes are neglected), when the state of the frame can be represented by fig. 9.2(d), the axial load in the columns is above the deteriorated stability value of $0.138P_e$. In this case, immediate collapse would occur on the formation of this third hinge, and the point R' in fig. 9.3 would represent the effective (and unstable) collapse load.

In this particular numerical example, the elastic critical load after the formation of the penultimate hinge was less than a quarter of the critical load for the original frame. Much more dramatic drops can occur for multi-storey frames as the formation of successive hinges makes the structure more and more flexible. Thus there would seem to be some serious difficulties in the way of extending simple plastic theory to the design of tall building frames. These difficulties may well be of the designer's own making, however, in the sense that the simple conceptual model on which design is based may be *too* simple to give a balanced view. Leaving out of account the question of strain hardening, which in itself may go a long way to alleviate the apparent difficulties, there are two other aspects of the design process which repay rather closer examination.

In the first place, all the calculations have been made, as is conventional, for a bare framework. While it may be true that certain types of cladding add very little strength to a building (for some types of cladding this is patently not true), almost all cladding will increase the stiffness of the whole structure. Thus almost certainly the '$P\Delta$' effects will be very much less than those calculated, that is, deflexions of the completed building will be much smaller than those calculated for the bare frames. Further, it is precisely stiffness, and not strength, that is required to resist tendencies to instability. (There are, indeed, unusual industrial buildings with open frameworks and no walls, but they are very unusual if they also have significantly high axial loads which might make frame instability a real problem.)

Secondly, the function of the load factor must once again be examined closely (further discussion in another context is given in section 9.3 below). A frame will usually remain elastic under the *working* values of the loads; one or two hinges might be formed in a complex structure, but these will not affect in any dramatic way the value of the elastic critical load in the working state. Thus if plastic methods of design are viewed (as they have tended to be viewed in these two volumes; cf. the remarks in section 6.8 earlier) as methods of deriving reasonable equilibrium states under working loads, there may be no question of any deterioration of stability. The question is whether the calculated load factor is regarded as a *hypothetical* increase in the value of the loads, or whether it represents some statistical likelihood of the loads actually reaching their collapse values; even in the second case, some at least of the margin of safety must be allowed for all the other imperfections of the design process.

In section 9.4 later, an example is given (for the braced frame) of a situation which is apparently less critical if plastic hinges are formed; calculations performed for working values of the loads lead to a *smaller* estimate of the margin of safety than do calculations in the hypothetical collapse state. For the unbraced frame the situation is reversed, since the formation of plastic hinges does lead to a deterioration in the overall stability of the frame.

Bearing all these considerations in mind, it would seem that a satisfactory model for design of tall buildings is one in which the stiffening effects both of cladding and of strain hardening are ignored; calculations are then made by simple plastic theory (in so far as strength is *prima facie* the primary design criterion) for the bare frames. Further, the design method may be viewed as one which establishes working conditions for the frames, and a margin of safety is provided by a suitable choice of load factor. Unless the frame is unusual, deterioration of stability, due to the formation of plastic hinges at load factors greater than unity, is unlikely to prove troublesome; naturally, some stability calculations may have to be made. Similarly, the effect of deflexions may not be nearly so marked as calculations for bare frames might imply. For tall unbraced frames deflexions may be of overriding importance, but the conclusion to be drawn here is that the design concept is bad; every attempt should be made to brace a tall frame against wind if wind loads start to govern the whole design.

9.2 Braced frames (1)

No discussion will be made here of the design of the bracing itself, whose function is to carry the wind load down to the foundations of the building. It will be assumed that there is sufficient bracing in the plane of the frames themselves, for example, or that the individual floors carry the wind loads to a braced central core of the building, so that the beams and columns are to be designed only for vertical dead and superimposed loading.

With the essential requirement that instability criteria cannot be used for design, there is considerable practical agreement that an effective way of designing multi-storey frames is to use 'weak' beams and 'strong' columns. That is, beams can be designed to have their smallest possible sections under the specified loading, but columns should have sufficient margin to ensure that there is no danger of instability. This requirement does not in fact rule out the possibility of the development of plastic hinges in columns, particularly for low buildings in which axial loads are small, provided that it can be demonstrated that the development of plasticity is a stable process. However, the implication for the general multi-story frame is that the columns should probably remain elastic under the factored vertical loads, that is, at collapse of the beams. (This last statement presupposes a certain way of design, based on the collapse state of the frame, which is open to criticism; see section 9.4 below.)

Accepting for the moment the idea of weak beams and strong columns, the design of the beams themselves is straightforward. As has been mentioned, there is no question of alternative loading patterns; each beam can be designed to collapse with hinges at its ends and centre. This plastic design of the beams is usually direct, but on infrequent occasions it may be necessary to make iterative calculations. For example, an outside column may prove on later examination to have inadequate section to carry the full plastic moment of one of the adjacent beams. In general, however, for spans and loadings of roughly comparable magnitudes, the full plastic moment at the end of a beam can be developed by the other members framing into that end, so that each span can usually be designed in isolation.

By contrast, the design of elastic columns in a multi-storey frame is always a two-stage iterative process. First, the bending moments must be determined for the particular column under consideration, and then a section must be assigned to the column to carry those moments together with the axial load that is acting. Since the one calculation affects the other, any attempt at a general solution is prohibitively complex. However, the work can be simplified considerably if use is made of the substitute frame illustrated in fig. 9.4. Here, the column being designed is regarded as connected to the adjacent members, but there is no further 'spread' into the structure as a whole.

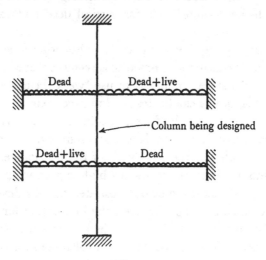

Fig. 9.4

Thus fig. 9.4 represents a small portion of a plane multi-storey multi-bay frame, for which it is assumed that remote members and loads have no effect on the column under consideration. In fig. 9.4 the outer ends of all the members are shown as fixed against rotation, but other end conditions may be used if these seem more suitable for a particular problem. For simplicity, it will be assumed in what follows that the column in fig. 9.4 is being bent about its major axis by the four beams to which it is connected. A view at right angles would show the column connected to 'minor-axis' beams (which may or may not be loaded, and, indeed, which might be omitted); the following remarks, although phrased in terms of major-axis bending, should be thought of as applying also to the minor-axis view. The two situations can then be combined for the final stability test of the column.

It will be noted that some live load has been omitted in fig. 9.4. This arrangement will tend to bend the central column length into single curvature, which is generally the most unfavourable configuration for stability. Under the full factored dead plus live loading two of the beams will be on the point of collapse, and the state of the substitute frame of fig. 9.4 will be that shown in fig. 9.5 (*a*). The problem now is to determine the bending moments in the central column length under these conditions. This is a fairly simple problem in elastic analysis; if necessary, the reduced stiffness of the column due to axial load can be taken into account. The stiffnesses of the collapsing beams are effectively zero, since they cannot absorb any further bending moment. The reduced substitute frame of fig. 9.5 (*b*) is therefore used for making the stability check on the column; the collapsing beams have been replaced by 'dead' moments of value M_p.

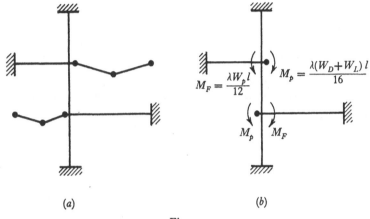

$$M_F = \frac{\lambda W_p l}{12} \qquad M_p = \frac{\lambda (W_D + W_L) l}{16}$$

(*a*) (*b*)

Fig. 9.5

(The two values of M_p marked in fig. 9.5 (*b*) are not, of course, necessarily equal. The fixed-end moments M_F correspond to a uniformly distributed load, as does the marked value of M_p, but the appropriate values should be used for more complex loading systems.)

The out-of-balance bending moments in fig. 9.5 (*b*) can be used as starting values for a moment-distribution process, or some other method may be used to give the values of the end bending moments acting on the central column length. These values, together with that of the (factored) axial load, can then be used to check the suitability of the chosen column section according to any required criterion (e.g. that of stability, perhaps together with the condition that the column should just remain elastic).

As a numerical example, which is based upon an actual design,† conditions for the first-floor column length in fig. 9.6 will be calculated;

† J. Heyman, J. C. H. Finlinson and R. P. Johnson, Inglis A: a fully rigid multi-storey welded steel frame, *The Structural Engineer*, **44**, 435–41, 1966.

note that the ground-floor column is assumed to be pinned to the foundations. The four beams are identical, and each is subjected to uniformly
distributed loads of 600 kN dead and 600 kN live. The load factor to be
used is 1·5 (this value is reasonable for a braced frame of this type).†
Thus the full plastic moment required for the beams is given by

$$M_p = (1·5)(1200)(16)/(16) = 1800\,\text{kN m};$$

this value is shown in fig. 9.7. The dead loads on the elastic beams will
produce (factored) fixed-end moments of

$$(1·5)(600)(16)/(12) = 1200\,\text{kN m}.$$

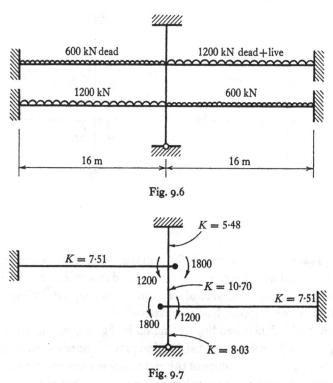

Fig. 9.6

Fig. 9.7

Thus an out-of-balance bending moment of value 600 kN m acts at the
top and bottom of the column length being considered.

The stiffnesses K marked in fig. 9.7 correspond to actual sections
having the required value of M_p for the beams, and to trial sections for
the columns. Using a method of moment distribution, the end conditions

† *Fully rigid multi-storey welded steel frames*, Joint Report of the Institution of
Structural Engineers and the Institute of Welding, December 1964.

on the centre column length are found to be those shown in fig. 9.8(a); the value of the axial thrust is determined by summing the floor loads for the upper storeys in the usual way.

Two courses are now open to the designer, depending on the design criterion he is using for the columns. The *collapse* values of fig. 9.8(a), together with the factored value of the axial thrust, could be used to check that the column remains (for example) both stable and elastic. In this case, the criterion being used for the column is a 'limit state' criterion, with no internal margin of safety; the overall safety of the structure is assured by the *premultiplication* of the loads by the load factor (1.5 in this example).

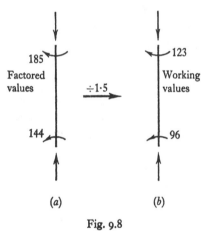

Fig. 9.8

Alternatively, the designer can view the elastic–plastic method he has used as one which generates reasonable values of the required quantities under the *working* values of the loads. In this case, the numbers shown in fig. 9.8(b) represent values of the bending moments in equilibrium with the applied loads under unit load factor. (Figures 9.8(a) and (b) are related simply by the factor 1·5.) The column could then be checked against some different criterion (for example, one given in a conventional elastic design code) which contains its own built-in margin of safety. (It is not necessarily illogical to use different load factors for different portions of a structure. The plastic collapse of beams is probably definable to closer statistical limits than is the elastic–plastic buckling of columns. It would be reasonable then to use a load factor of 1·8 (say) on the columns as against 1·5 on the beams.)

9.3 The load factor

There are several criticisms which can be made of the design process which has just been outlined. In the first place, it leans heavily on the conventional idea that, if the end conditions of a single column length can be determined, then this gives sufficient information to check the stability of the column. The idea that this might be possible springs from the concept of the single-length pin-ended column on which so many early tests were done. Later work was, it is true, more sophisticated in that experiments were made with eccentric loading and, in some cases, with separately-applied end moments. Even so, it is not obvious that these results are immediately applicable to the actual problem of a continuous column running through several storeys of a building frame. In fact, the failure of such a continuous column length can involve simultaneous buckling in more than one storey, and the behaviour is much influenced by the stiffnesses of secondary members framing into the column. Further, the bending moments of fig. 9.8 should not really be regarded as having fixed values; these will alter as the column deforms.

However, design on the basis of a single column length is very probably conservative, at least in the presence of secondary restraining members whose stiffnesses would be ignored by any simple theory. Moreover, the full problem, of accounting for the behaviour of a continuous column connected about both axes at each storey height to beams which can apply bending to the column, is one of very great complexity. Fortunately, a conservative approach to column design is not usually unduly wasteful of material. The calculations will appear to be very sensitive to the value of slenderness ratio for the column, but it will often be found that the abandonment of one class of column, say 10×10 Universal Columns, in favour of the next higher class (12×12 in this case) will enable a column design criterion to be satisfied without necessarily increasing the weight of the section used.

Accepting for the moment, then, the idea of single-length column design (or, really, column checking), there are still some anomalies in the steps which led from fig. 9.6 to fig. 9.7 to the design values of fig. 9.8. It is clear, for example, that more severe conditions would arise for the central column length in fig. 9.5 (b) if the dead-load moments M_F, which oppose the full plastic moments M_p, were not factored. The use of an overall load factor (of value 1·5 in this numerical example) on both dead and live loading allows the dead load to relieve partially the bending moments in the column. In such cases, it might be appropriate to use a load factor of unity (or perhaps 0·9) on the dead load. If a factor of unity were used for the numerical example of fig. 9.6, then fig. 9.7 would be replaced by fig. 9.9(a), and the resulting bending moments on the central column length (fig. 9.8) would be modified to those of fig. 9.9(b) and (c). Some increase (in the ratio 7/6) is seen in the values of the bending moments.

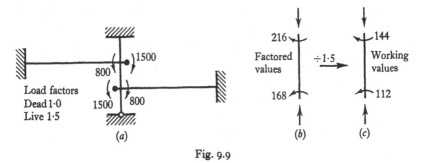

Fig. 9.9

This last analysis, even though it led to higher values for the bending moments, would even so be rather dangerous. It will be seen that the full plastic moment of the beams in fig. 9.9(a) has been reduced to 1500 kN m from the value of 1800 in fig. 9.7. This reduction has come about because the beams are supposed to be collapsing under a total load of

$$600 + (1·5)(600) = 1500 \, \text{kN}$$

instead of a load of $(1·5)(600 + 600) = 1800$ kN. Now the load factor of 1·5 used in these calculations is, from one point of view, no more than a simple number introduced to lead to final designs which will prove satisfactory in practice. It may well be the case that an *overall* factor of 1·5 applied to both dead and superimposed loading will lead to designs of braced frames (of the type considered here) which are indeed satisfactory. If this is so, however, then it is clearly dangerous to attempt to draw 'logical' conclusions from some pseudo-rational interpretations of the function of the load factor—to say, for example, that since the dead load is known much more accurately than the live load, then the dead-load factor can be reduced to unity while the live-load factor *remains at* 1·5.

Design with different load factors is certainly possible. For the example above, with the particular ratio of dead to live load on the beams (in fact, equal values), a dead-load factor of unity combined with a live-load factor of 2·0 would be satisfactory; the combined effective load factor against collapse of the beams would be 1·5. In this case, the values of the full plastic moments in fig. 9.9(a) would be set back to 1800 kN m, the elastic fixed-end moments remaining at 800 kN m; the resulting bending moments in figs. 9.9(b) and (c) would be further increased by a factor of 10/7.

The kind of juggling with numbers indulged in in the last paragraph can be extended in much more tortuous ways. Any required value of the load factor can be built up by assigning appropriate partial safety factors to each of the aspects of design, bearing in mind the scatter of material properties, statistical patterns of loading, standards of workmanship, possible danger to life in case of collapse, and so on. In whatever way these numbers are determined, however, it should always be arranged that the combined overall load factor corresponds to a value that has been found, in practice, to give safe designs; indeed, the relative importance of the various partial factors should be fixed so that the 'right' answer is obtained.

Once a value for the load factor has been settled, then all the ideas of simple plastic theory discussed in this and in the previous volume can be brought into the design process. The idea of working loads gradually increasing in proportion leads to the simple but very powerful notion of the plastic collapse of beams, and to the development of the master theorems which deal with the overall safety of a frame. The fact remains that the concept of a collapse load factor is reflected only in an insignificantly small probability of an actual overload of a real structure in practice; the value of load factor to be used in design is one which has been proved, by experience, to give satisfactory structures of the class being considered.

9.4 Braced frames (2)

From the last paragraph, it may be concluded that the limited substitute frame of fig. 9.4 will, for all practical purposes, *never* reach the state illustrated in fig. 9.5(a), in which the two loaded beams are on the point of collapse. It is the behaviour of the frame under the nominal working loads which is of interest. Thus, in the numerical example of fig. 9.6, the full plastic moment of the beams may be determined quite properly to have the value 1800 kN m, but this does not mean that a plastic mechanism of collapse, as in fig. 9.5(a), should be used to determine the most critical conditions for the columns. Rather more severe conditions for the columns than those shown in fig. 9.8 will be obtained if the calculations are made for working rather than factored values of the loads.

The working loads shown in fig. 9.6 will give the fixed-end moments marked in fig. 9.10(*a*); note that the full plastic moments of the beams are all 1800 kN m, so that the value of 1600 kN m shown is in fact elastic. As before, the out-of-balance bending moments, 800 kN m at each end of the column, may be apportioned between the members by the method of moment distribution; the stiffness of the *whole* substitute frame of fig. 9.10(*a*) must be considered, since plastic hinges do not 'cut off' any of the beams. The resulting end bending moments on the central column length are shown in fig. 9.10(*b*); it will be seen that these exceed by some 40 per cent the working values marked in fig. 9.8(*b*). Similarly, the factored values marked in fig. 9.10(*c*), obtained simply by applying a (hypothetical) load factor of 1.5, may be compared with the factored values marked in fig. 9.8(*a*).

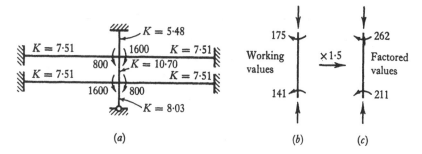

(*a*)   (*b*)   (*c*)

Fig. 9.10

Thus, however the column length is checked, whether on a working-load or on a factored basis, it is clear that the conditions of fig. 9.10(*b*) and (*c*) are more severe than those of fig. 9.8. The apparent margin of safety of the column depends on the conceptual model of the structure as a whole. This situation seems unsatisfactory, at least when related to the more or less definite answers afforded by conventional design rules. The fact that an answer is definite does not necesssarily mean that it is correct, however. It is perhaps one of the virtues of plastic methods that, while not in all cases supplying complete answers, it does at least pose a different set of questions to be resolved by the designer. In the present case, with inadequate theoretical studies of the buckling of continuous columns on the one hand, and insufficient practical experience to determine minimum values of the load factor on the other, the designer is forced to use his judgment in choosing between figs. 9.8 and 9.10. If there are no external guides whatever in this choice, then presumably the worse situation must be taken.

9.5 Sway frames

The 'strong-column weak-beam' concept will be used to construct a very crude model, from which preliminary ideas can be obtained of the effect of wind on unbraced frames. A sketch design will be made of the frame of fig. 9.11; despite the crudity of the model, it will be found that certain patterns of behaviour can be established. The frame has m bays, each of span l, and n storeys, each of height h. Loading will be taken as uniform and identical on all floors; if the total load per unit area is w, then the loads W in fig. 9.11 are given by (for frames spaced L apart):

$$W = wlL. \tag{9.2}$$

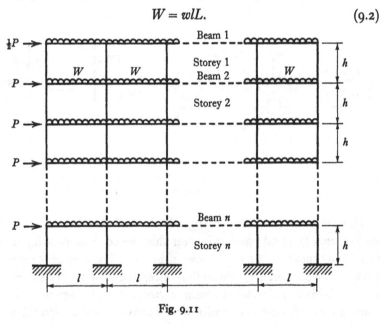

Fig. 9.11

The loads P in fig. 9.11 represent the effect of wind loading, and are taken to act at beam level; if the unit wind pressure (specified by the Codes) is p, then

$$P = phL. \tag{9.3}$$

The storeys are numbered from the *top*, as shown. For storey r, the total wind load carried is

$$\tfrac{1}{2}P + P(r-1) = \tfrac{1}{2}P(2r-1). \tag{9.4}$$

Some numerical calculations will be made, using the following values:

$$w = 8\,\text{kN/m}^2, \quad p = 2 \cdot 5\,\text{kN/m}^2, \atop l = 10\,\text{m}, \quad h = 4\,\text{m}, \quad L = 6\,\text{m},} \tag{9.5}$$

from which

$$\left.\begin{aligned} W &= 480\,\text{kN}, \\ P &= 60\,\text{kN}, \\ Wl &= 4800\,\text{kN m}, \\ Ph/Wl &= 0 \cdot 05. \end{aligned}\right\} \tag{9.6}$$

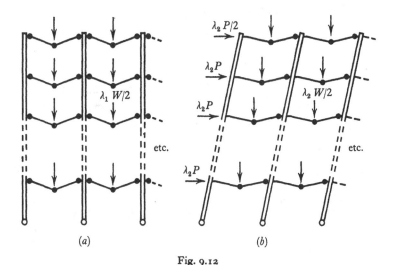

(a)   (b)

Fig. 9.12

Load factors of value 1·75 will be used for the vertical loading, and 1·40 for combined vertical plus wind loading. For tall sway frames whose design is really dictated by the wind the factor of 1·40 may well be inadequate. However, as will be seen immediately, the use of values such as 1·75 and 1·40 helps to give the designer some idea of when wind becomes important.

For the purpose of the crude analysis, the frame of fig. 9.11 is replaced by that of fig. 9.12. The columns of the frame are supposed to be pinned at their feet, instead of fixed, and to be so strong that they remain effectively rigid during the formation of a collapse mechanism. The distributed loads W of fig. 9.11 have been replaced by concentrated loads $\frac{1}{2}W$; the unfactored free bending moment at mid-span of each beam has value $\frac{1}{8}Wl$ in either case.

9. MULTI-STOREY FRAMES

Under vertical load alone each beam must collapse independently by the development of hinges at its ends and centre, fig. 9.12(a). If the full plastic moment of the beam is B_0, then the design of the frame is given by

$$B_0 = \lambda_1(\tfrac{1}{16}Wl). \tag{9.7}$$

Figure 9.12(b) shows the collapse mechanism of this frame under the action of vertical and wind loads combined. If the rotation of each (rigid) column is θ, then the work done by the vertical loads in each span is $\lambda_2(\tfrac{1}{2}W).(\tfrac{1}{2}l)\theta$, while the topmost horizontal load does work $\lambda_2(\tfrac{1}{2}P)nh\theta$, the next $\lambda_2 P(n-1)h\theta$, and so on. Since the total hinge rotation in each beam is 4θ, the collapse equation is

$$\lambda_2[mn(\tfrac{1}{4}Wl) + Ph\{\tfrac{1}{2}n + (n-1) + (n-2) + \ldots + 1\}] = 4B_0 mn,$$

that is

$$\lambda_2(mWl + 2nPh) = 16B_0 m. \tag{9.8}$$

The collapse mechanism in fig. 9.12(b) can form as assumed provided the bending moment at the left-hand end of each beam remains less than B_0; a simple statical analysis gives as this condition

$$\frac{n}{m} < \frac{1}{2}\frac{Wl}{Ph}. \tag{9.9}$$

Thus for given loads the number n of storeys for which the assumed collapse mechanism can occur is limited; as will be seen below, however, a more stringent condition than (9.9) is imposed on the number of storeys.

For any given values of λ_1 and λ_2, equations (9.7) and (9.8) serve to give an estimate of the critical height of a frame for which wind loads became the over-riding design criterion. Thus if the design of the frame is made according to equation (9.7), then that frame will collapse under wind loading at a load factor λ_2 given from equation (9.8) by

$$\lambda_2(mWl + 2nPh) = \lambda_1 mWl,$$

that is

$$\frac{n}{m} = \left(\frac{\lambda_1}{\lambda_2} - 1\right)\left(\frac{1}{2}\frac{Wl}{Ph}\right). \tag{9.10}$$

As a numerical example, setting $\lambda_1/\lambda_2 = 1\cdot75/1\cdot40 = 1\cdot25$, and using values (9.6), equation (9.10) gives $n/m = 2\cdot5$. Thus if a single-bay frame ($m = 1$) is designed for vertical load only according to equation (9.7), it will collapse under wind at a load factor lower than $1\cdot4$ if the number of storeys is three or more. Similarly, a ten-bay frame is limited to twenty-five storeys.

If a frame has a number of storeys greater than the limiting value given by equation (9.10), then wind load becomes the design criterion and the beam sections must be increased from the values given by (9.7), in order that the load factor λ_2 should be maintained at an acceptable value. This increase in beam section can be applied to all the beams; from equation (9.8), each beam would have full plastic moment

$$B_0 = \lambda_2 \frac{Wl}{16}\left(1 + \frac{2n}{m}\frac{Ph}{Wl}\right). \tag{9.11}$$

Alternatively, each beam could be of different section, full plastic moment B_r at storey r, in which case equation (9.11) can be rewritten:

$$\sum_{r=1}^{n} B_r = \lambda_2 \frac{Wl}{16}n\left(1 + \frac{2n}{m}\frac{Ph}{Wl}\right). \tag{9.12}$$

If the designs of an $(r-1)$-storey and an r-storey frame are identical apart from the addition of the extra beam for the taller frame, then equation (9.12) gives

$$B_r = \lambda_2 \frac{Wl}{16}\left[1 + 2\left(\frac{2r-1}{m}\right)\frac{Ph}{Wl}\right]. \tag{9.13}$$

Thus *one* way of designing a frame of height greater than that implied by equation (9.10) would be to assign sections to the upper storeys with full plastic moment $B_0 = \lambda_1 Wl/16$, and to start increasing this beam section when the limit of equation (9.10) is passed. As an example, a single-bay frame with the loads of (9.6) would have the design:

$$\left.\begin{array}{l} B_0 = \lambda_1(\tfrac{1}{16}Wl) = 525\,\text{kN m}, \\[4pt] B_2 = B_1 = B_0, \\[4pt] B_3 = (3\cdot12-2)\,B_0 = 588\,\text{kN m}, \\[4pt] B_4 = 1\cdot36 B_0 = 714\,\text{kN m}, \\[4pt] B_5 = 1\cdot52 B_0 = 798\,\text{kN m}, \end{array}\right\} \tag{9.14}$$

increasing thereafter by $0\cdot16 B_0 = 84\,\text{kN m}$ for each storey added.

The use of point loads rather than distributed loads affects these conclusions only slightly. Using the full distributed value of W, equation (9.7) remains unchanged, while equation (9.8) becomes

$$\lambda_2 Wl\left(1 + \frac{n}{m}\frac{Ph}{Wl}\right)^2 = 16 B_0. \tag{9.15}$$

The only difficulty in deriving equation (9.15) lies in the determination of the position of the sagging hinge in each span. Equation (9.10) giving the limiting number of storeys becomes

$$\frac{n}{m} = \left(\sqrt{\frac{\lambda_1}{\lambda_2}} - 1 \right) \frac{Wl}{Ph};$$ (9.16)

using the previous numerical example, $n/m = 2 \cdot 36$ instead of the value $2 \cdot 5$ derived by using point loads.

Having made a preliminary study in this way of the behaviour of an unbraced frame, the designer can then make more detailed calculations which can be used as a basis for the final design of a particular frame. In particular, the bending moments in the columns must be estimated, so that stability checks can be made. For low unbraced buildings, where the effect of wind is small, the problem is very similar to that discussed in sections 9.2 and 9.4 above for the braced frame; patterns of loading must be considered in order to estimate worst conditions for the columns. For tall frames, on the other hand, the 'sway moments' due to wind are likely to lead to the largest values of bending moment in the columns.

Under sway conditions, the final checking of the columns should probably be done using the results of an elastic or of a plastic-collapse computer analysis. The same difficulties arise in connexion with the interpretation of the load factor that were mentioned in section 9.3; that is, an (elastic) working-load analysis can be used to make a check on the columns, or a collapse analysis can be 'scaled down' by the load factor to give a similar estimate of working-load conditions. In either case, the method is one of the analysis of a *given* frame; what is required first is some method for assigning trial sections to the columns.

For this purpose various arrangements of plastic hinges can be inspected in order to determine equilibrium distributions of bending moment. Such arrangements of hinges could include hypothetical hinges in the columns; the final design of the column might assign sections such that the columns remained both stable and elastic. Thus the steps in a possible design method for a weak-beam strong-column frame are as follows:

(i) A system of plastic hinges is assumed, involving collapse in both beams and columns.

(ii) From the assumed hinge locations, relations may be established from which the full plastic moments of the beams and columns can be calculated.

(iii) The sections of the beams are chosen so that their full plastic moments correspond to those calculated in step (ii); that is, the beams will be just strong enough for the assumed collapse mechanism to develop when the (factored) loads are applied to the frame.

(iv) Using the values of full plastic moment for the columns derived in step (ii), together with the known axial thrusts, stability checks can be made on individual column lengths.

It will be appreciated that this method of design is *safe* according to the principles of simple plastic theory; in particular, the provision of columns stronger than those necessary theoretically cannot weaken the frame, provided there is no danger of instability and that deflexions remain small.

A possible pattern of hinges, on which a design may be based, is shown in fig. 9.13; the distributed loads have again been replaced for the time

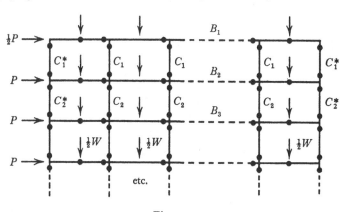

Fig. 9.13

being by concentrated loads of value $\frac{1}{2}W$. (There are many other patterns possible. In particular, Tanabashi and Nakamura† have given hinge patterns for the formation of a Foulkes mechanism corresponding to the minimum-weight design of multi-storey frames. However, the arrangement of hinges in fig. 9.13 leads to reasonably simple expressions for the values of full plastic moment.)

In fig. 9.13, it is assumed that all *internal* columns in one storey have the same full plastic moment, those in storey r having full plastic moment C_r, while the two external columns in the same storey have full plastic moment C_r^*. It will be seen that each storey can sway independently, and, while material could be transferred from internal to external columns, the full plastic moments are the least possible for each storey. This

† R. Tanabashi and T. Nakamura, The minimum weight design of a class of tall multi-storey frames subjected to large lateral forces, *Proc. fifteenth Japan National Congress for Applied Mechanics*, Tokyo, 1965.

assumed pattern of hinges for the columns corresponds, of course, to taking points of contraflexure at the centre of each column. However, the plastic approach would *force* the points of contraflexure to stay at the column centres at collapse if the column sections were chosen to be only just strong enough (i.e. a weak-column design).

The shear force on each internal column is $2C_r/h$; using expression (9.4) for the total wind shear on a storey,

$$(m-1)\frac{2C_r}{h}+2\frac{2C_r^*}{h}=\frac{P}{2}(2r-1),$$

that is
$$(m-1)C_r+2C_r^*=\frac{Ph}{4}(2r-1). \tag{9.17}$$

Similarly,
$$(m-1)C_{r-1}+2C_{r-1}^*=\frac{Ph}{4}(2r-3), \tag{9.18}$$

and, denoting
$$\left.\begin{array}{l}X_r=C_r+C_{r-1},\\ X_r^*=C_r^*+C_{r-1}^*,\end{array}\right\} \tag{9.19}$$

equations (9.17) and (9.18) may be added to give

$$(m-1)X_r+2X_r^*=Ph(r-1). \tag{9.20}$$

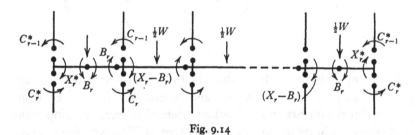

Fig. 9.14

For the top storey, $r=1$, equation (9.20) should be replaced by

$$(m-1)X_1+2X_1^*=\tfrac{1}{4}Ph, \tag{9.21}$$

as is evident from equation (9.17).

A typical beam r is sketched in fig. 9.14; the beam is assumed to have constant full plastic moment B_r. Independent collapse occurs in the left- and right-hand spans, while the internal spans each have a single plastic hinge. From the collapse conditions for the left-hand span,

$$B_r=\tfrac{1}{12}Wl+\tfrac{1}{3}X_r^*, \tag{9.22}$$

while from the right-hand span,

$$B_r = \tfrac{1}{12}Wl + \tfrac{1}{3}X_r - \tfrac{1}{3}X_r^*. \tag{9.23}$$

Equations (9.22) and (9.23) combine to give

$$X_r = 2X_r^*, \tag{9.24}$$

(that is, the internal columns have twice the strength of the external columns) and substitution into equation (9.20) gives

$$X_r = 2X_r^* = \frac{Ph}{m}(r-1). \tag{9.25}$$

The design of the frame is thus given finally by

$$\left. \begin{aligned} X_r &= \frac{Ph}{m}(r-1), \\ X_r^* &= \frac{Ph}{2m}(r-1), \\ B_r &= \frac{Wl}{12} + \frac{Ph}{6m}(r-1). \end{aligned} \right\} \tag{9.26}$$

(Note that $X_1 = 2X_1^* = Ph/4m$, $B_1 = \tfrac{1}{12}Wl + Ph/24m$.)

This design is valid providing the bending moments at cross-sections in fig. 9.14 which are not hinge positions remain below the fully plastic values. For the left-hand span $X_r^* \leqslant B_r$, and for the right-hand span $(X_r - B_r) \leqslant B_r$. These inequalities are the same, by virtue of equation (9.24), and substitution of values (9.26) gives

$$\frac{r-1}{m} \leqslant \frac{1}{4}\frac{Wl}{Ph}. \tag{9.27}$$

For the previous numerical values, $(r-1) \leqslant 5m$, that is, the design is valid for a fifty-one-storey ten-bay frame.

The design for the actual problem of distributed loads assumes the same arrangement of hinges, figs. 9.13 and 9.14, but with the sagging hinges in the beams moved off-centre to the positions of maximum bending moment. Determination of the locations of these hinges leads to quadratic equations; equations (9.17) to (9.21) remain unchanged, but equations (9.22) and (9.23) are replaced by

$$B_r = \tfrac{3}{2}Wl - X_r^* - 2\left[2\left(\frac{Wl}{2}\right)\left(\frac{Wl}{2} - X_r^*\right)\right]^{\frac{1}{2}}, \tag{9.28}$$

and $$B_r = \tfrac{3}{2}Wl + X_r + X_r^* - 2\left[\frac{Wl}{2}\left(2\frac{Wl}{2} + X_r + 2X_r^*\right)\right]^{\frac{1}{2}}. \tag{9.29}$$

Using the substitution
$$Z_r = X_r + 2X_r^*, \tag{9.30}$$

equations (9.28) and (9.29) may be combined to give

$$\frac{(r-1)Ph - Z_r}{m-2} = \frac{1}{2}\frac{Z_r}{Wl}(Z_r + 4Wl) - 2\frac{Z_r}{Wl}\left[\frac{Wl}{2}(Z_r + Wl)\right]^{\frac{1}{2}} = X_r. \tag{9.31}$$

For any value r, equation (9.31) may be solved numerically for Z_r and hence for X_r; X_r^* may then be found from (9.30), and finally B_r from either (9.28) or (9.29).

The limit corresponding to inequality (9.27) is

$$\frac{r-1}{m} \leqslant \frac{1}{2}\frac{Wl}{Ph}; \tag{9.32}$$

using the same numerical values, the design for distributed loads is valid for a 101-storey ten-bay frame.

Slight modifications are required for single- and two-bay frames. For a two-bay frame, equation (9.31) indicates

$$Z_r = (r-1)Ph, \tag{9.33}$$

this being the only change. A separate analysis must be made for the single-bay frame; the results are

$$\left.\begin{aligned}
X_r^* &= \tfrac{1}{2}Ph(r-1), \\
[X_1^* &= \tfrac{1}{8}Ph], \\
B_r &= \tfrac{1}{8}Wl[1 + (Ph/Wl)^2\{2(r-1)\}^2], \\
[B_1 &= \tfrac{1}{8}Wl\{1 + \tfrac{1}{4}(Ph/Wl)^2\}],
\end{aligned}\right\} \tag{9.34}$$

and the analysis is valid for any height of frame.

As an example, the single-bay equations (9.34) give

$$\left.\begin{aligned}
X_1^* &= C_1^* = \tfrac{1}{8}Ph, \\
X_2^* &= \tfrac{1}{2}Ph, \quad C_2^* = \tfrac{3}{8}Ph, \\
X_3^* &= Ph, \quad C_3^* = \tfrac{5}{8}Ph, \\
X_4^* &= \tfrac{3}{2}Ph, \quad C_4^* = \tfrac{7}{8}Ph, \quad \text{etc.},
\end{aligned}\right\} \tag{9.35}$$

and

$$\left.\begin{aligned}
B_1 &= \tfrac{1}{8}Wl[1 + \tfrac{1}{4}(Ph/Wl)^2], \\
B_2 &= \tfrac{1}{8}Wl[1 + 4(Ph/Wl)^2], \\
B_3 &= \tfrac{1}{8}Wl[1 + 16(Ph/Wl)^2], \\
B_4 &= \tfrac{1}{8}Wl[1 + 36(Ph/Wl)^2], \quad \text{etc.}
\end{aligned}\right\} \tag{9.36}$$

Using the same numerical values (9.6), with a load factor of 1.4, equations (9.35) and (9.36) give

$$
\begin{rcases}
C_1^* = 42, & B_1 = 840, \\
C_2^* = 126, & B_2 = 848, \\
C_3^* = 210, & B_3 = 874, \\
C_4^* = 294, & B_4 = 916, \\
C_5^* = 378, & B_5 = 974.
\end{rcases}
\tag{9.37}
$$

These numerical values may be compared directly with those of (9.14) for the frame with very strong columns.

In practice, it is likely that a column having a full plastic moment of 200 to 300 kN m must be provided for stability reasons; thus the basic design of (9.37) could be modified in the top two or three storeys, with the beam sections possibly reduced slightly (although the requirement that the beams should carry vertical load only at a load factor of 1·75 will limit the amount of reduction possible).

All these results are really only illustrative of the way in which plastic theory permits a much more direct approach to the problem of design than do the trial-and-error analyses of elastic theory. A regular frame of the type illustrated in fig. 9.11, with equal storeys, equal bays, and equal floor loads, is never encountered in practice; however, patterns of hinges such as those of fig. 9.13 can be adapted very easily to the direct preliminary design of a practical frame. The design example of the next chapter shows that the crude idea of strong columns and weak beams can be used to generate a practicable structure.

The whole design process for multi-storey frames can, of course, be programmed for a computer; one such program† involves an initial selection of beam sizes according to plastic criteria, followed by an elastic–plastic analysis (allowing for the formation of plastic hinges) to determine the column bending moments. As usual, an iterative method is used for the design of the columns.

† K. I. Majid and D. Anderson, Elastic–plastic design of sway frames by computer, *Proc. Instn civ. Engrs*, 41, p. 705, 1968.

10

A DESIGN EXAMPLE

The frame to be discussed here is a slighly simplified version of a practical design for an electrical substation.† It is a requirement of the design that the interior of the building should be as free as possible from physical obstructions, and that electrical clearances for bus bars, isolators, circuit breakers and so on should be large. One way of constructing such a switch-house is to use unbraced frames, and calculations are given here for one such frame of the building.

The substation is intended as a standard building capable of being erected on poor foundations, and the column feet will therefore be taken as pinned. Similarly, a high unit wind pressure is taken corresponding to maximum exposure conditions. As will be seen below, although the frame has three bays and only two storeys, its height, 24 m, ensures that the wind forces play a major role in design. The frame carries very little dead load, the main gravity loading being due to equipment which can be removed periodically for servicing. The consequent high ratio of live to dead load, coupled with the relatively large wind forces, make the design sensitive to incremental collapse and alternating plasticity.

10.1 Frame dimensions and loadings

Figure 10.1 shows the centre-line dimensions of a typical frame for the switchhouse. The frame has pinned feet, and it will be assumed that connexions between the members are full strength. Such connexions can be made by site welding; alternatively, it is fairly easy to design bolted connexions, using high-strength bolts, which are fully capable of transmitting the required full plastic moments from one member of the frame to another.

The joints of the frame are numbered, in accordance with the requirements of the STRESS program‡ for the elastic analysis of frames on an IBM 1130 computer, from 1 to 33. A 'joint' is taken at every section of

† J. Heyman, R. P. Johnson, P. P. Fowler and I. P. Gillson, Shakedown analysis: The design of a 275 kV switchhouse, *The Structural Engineer*, 46, no. 4, p. 97, 1968.

‡ S. J. Fenves, R. D. Logcher, S. P. Mauch, and K. F. Reinschmidt, *STRESS: A user's manual*, Cambridge, Mass. (M.I.T. Press) 1964.

the frame where members meet or where loads can be applied. Thus the roof beam has a 'joint' at every 4 m, corresponding to the spacing of the roof purlins. The members between joints are also numbered, from 1 to 35. A particular section of the frame can be referred to by two numbers, given in the order member/joint; thus the reference 29/26 is to the left-hand end of the central span of the roof beam.

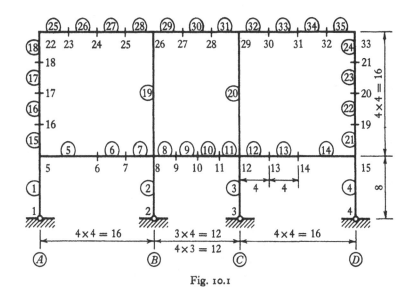

Fig. 10.1

The following loadings (illustrated in fig. 10.2; all loads in kN) were considered:

1. Dead load.

2. Superimposed load on the roof.

3. Equipment on left-hand bay.

3R. Equipment on right-hand bay.

4. Equipment on centre bay.

5. Wind load, wind from left to right.

5R. Wind load, wind from right to left.

The dead loads are due mainly to the self-weight of the steelwork, and allow for auxiliary members not considered here. These loads always act on the frame. Loading 3R is not illustrated in fig. 10.2; it is a mirror image of loading 3. It should be noted that loadings 3 and 3R can occur simultaneously, and that both are independent of loading case 4. Similarly, loading case 5R is a mirror image of case 5, but these two cases are, of course, alternative.

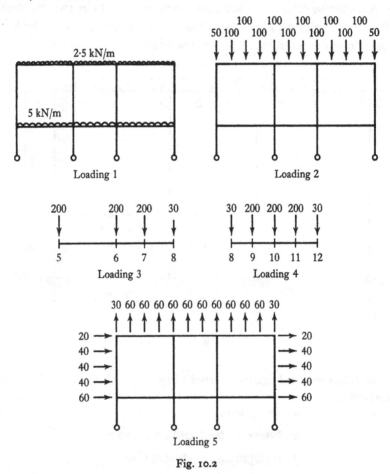

Fig. 10.2

10.2 Preliminary design: static collapse

Calculations will be made using *working* values of the loads, that is, using the values marked in fig. 10.2. Actual sections will be assigned to the members to provide a static collapse load factor of 1·75 against vertical loading, and of 1·40 against combined vertical and wind loading. Since the worst case for static plastic design occurs when all the gravity loads act together, loading cases 1 to 4 (including 3R) are superimposed to give the design loading of fig. 10.3 (*a*). The uniform loading of case 1, fig. 10.2, has been replaced by equivalent point loads at the 'joints'; loads acting at connexions between beams and columns affect only the axial loads in the columns, and hence have been omitted from fig. 10.3 (a).

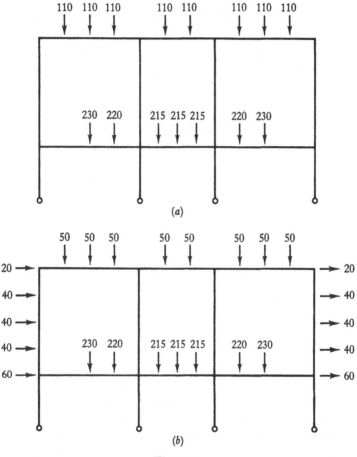

Fig. 10.3

The combined vertical and wind loading is illustrated in fig. 10.3 (b), and is determined merely by superimposing case 5 of fig. 10.2 and the loadings of fig. 10.3 (a). Thus the two loading cases of fig. 10.3 must be used for the static plastic design, the first to be taken with a load factor of 1·75, and the second with a load factor of 1·40.

Free bending moments may be determined conveniently by considering the frame of fig. 10.4. Frictionless pins have been inserted into the original frame so that every beam is simply supported, and hence statically determinate. Similarly, apart from the right-hand column, each column length is pin-ended. The right-hand column is continuous through both storeys, and the actual pinned foot has been replaced by a fixed end. The *vertical* loads of fig. 10.3 (*a*) and (*b*) will thus give rise to calculable bending moments in the beams of fig. 10.4, and to zero moments in the columns. The horizontal wind loads of fig. 10.3 (*b*) cause zero bending moments in the beams, and are carried by the right-hand column only. (This simple scheme of a statically determinate frame can be used for the general multi-storey multi-bay structure. All beams and column lengths are isolated with frictionless pins, with the exception of the right-hand column which extends continuously for the full height of the frame. If the frame has fixed feet, then all except that of the right-hand column are furnished with pins.)

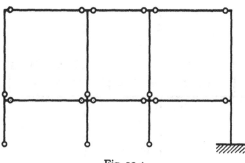

Fig. 10.4

The design will be started by assuming that vertical loading, with a load factor 1·75, will govern the design. Further, in accordance with the principles of the last chapter, final column sections will be chosen so that the columns remain elastic at the collapse of the frame as a whole. Thus plastic hinges are confined to the beams. If a uniform section is chosen for the roof beam, then it is clear from fig. 10.3 (*a*) that the outer spans will govern the design; fig. 10.5 (*a*) shows the left-hand span of the roof beam in isolation, and the free bending moments, for both ends pinned as in fig. 10.4, are given in table 10.1. The symmetrical collapse mode of fig. 10.5 (*b*) is critical, but the mode of fig. 10.5 (*c*) is also examined, since this will be required when considering the wind loading. Using the hinge rotations marked in fig. 10.5 in conjunction with the table of free bending moments, table 10.1:

$$\left.\begin{array}{l}\text{Fig. 10.5}(b)\text{:}\quad (-880)(-2) = 4M_p,\ M_p = 440\,\text{kN m.}\\[4pt]\text{Fig. 10.5}(c)\text{:}\quad (-660)(-\tfrac{4}{3}) = \tfrac{8}{3}M_p,\ M_p = 330\,\text{kN m.}\end{array}\right\}\quad(10.1)$$

Thus a section having a full plastic moment of 770 kN m must be provided in order to achieve a load factor of 1·75. If steel having a yield stress of 250 N/mm² is used, then the corresponding plastic section modulus is 3·08 × 10⁶ mm³; a 24 × 9 UB 113 kg/m has a quoted modulus of

$$3·283 \times 10^6 \text{ mm}^3,$$

and a corresponding full plastic moment of value 821 kN m. Thus the roof beam would have a load factor against collapse given by

$$\lambda = \tfrac{821}{440} = 1·87. \tag{10.2}$$

Table 10.1

Joint	22	23	24	25	26
Free B.M. (kN m)	0	−660	−880	−660	0

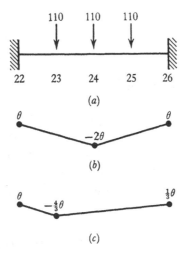

Fig. 10.5

The above calculation has neglected the effect of shear force on the value of full plastic moment. This effect is small, being about 3 per cent in the present case; the true load factor against collapse is about 1·81. Such effects will be ignored in all further calculations.

The free bending moments for the left-hand and central spans of the lower beam, under the loads of fig. 10.3(a), are given in table 10.2. The mechanisms of figs. 10.6(b) and 10.7(b) lead to the equations:

$$\left.\begin{array}{l} \text{Fig. 10.6}(b): \quad 2720 = 4M_p, \; M_p = 680 \text{ kN m.} \\ \text{Fig. 10.7}(b): \quad 2580 = 4M_p, \; M_p = 645 \text{ kN m.} \end{array}\right\} \tag{10.3}$$

273

Table 10.2

Joint	5	6	7	8	8	9	10	11	12
Free B.M. (kN m)	0	−1360	−1120	0	0	−967·5	−1290	−967·5	0

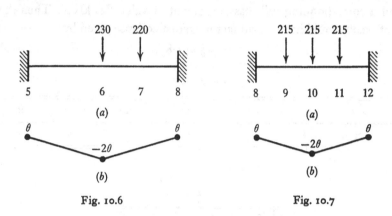

Fig. 10.6   Fig. 10.7

Each of the resulting values of M_p is of course exactly half of the corresponding maximum free bending moment in table 10.2; equations (10.3) will be used later on in the sway calculations. The two values of M_p are close, and the same section will be used throughout the lower beam length in this preliminary design. A 27×10 UB 152 kg/m has a full plastic moment of value 1247 kN m, again using steel with $\sigma_0 = 250$ N/mm². Thus the load factor against collapse of either of the outer spans under vertical load alone is

$$\lambda = \tfrac{1247}{680} = 1\cdot83. \qquad (10.4)$$

The two sections chosen so far are marked in fig. 10.8; the columns have not yet been designed, but their sections will be chosen finally so that they are strong enough for no plastic hinges to form in them. (It may be noted that both upper and lower beams have their minimum possible sections, and the columns will be assigned minimum sections according to a non-plastic limitation. The design point is thus at a vertex of a cube in the design space, and hence corresponds to a Foulkes mechanism of the type discussed in section 7.4 above; the design has minimum weight against vertical load alone.)

The frame of fig. 10.8 must now be analysed under the loading of fig. 10.3(b) to confirm the supposition that this wind loading case, at a load factor of 1·40, is less critical than vertical load only at a load factor of 1·75. The basic sway mechanism is shown in fig. 10.9. Comparison with fig. 10.4 shows at once that the free bending moments are zero at all the hinge positions of the sway mechanism, *with the exception of the real pin at the right-hand column foot*. The moment of the horizontal forces in fig. 10.3(*b*) about the right-hand column foot is 5760 kN m; with the usual

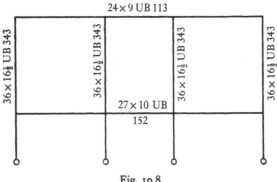

Fig. 10.8

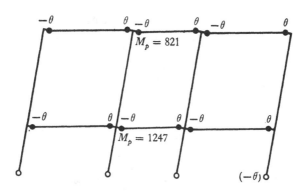

Fig. 10.9

sign convention, the free bending moment at the column foot has value -5760 kN m. Thus the virtual work equation for the loading of fig. 10.3(*b*) and the collapse mechanism of fig. 10.9 gives

$$(-5760\lambda)(-\theta) = (821)(6\theta) + (1247)(6\theta),$$

that is

$$5760\lambda = 12408; \quad \lambda = 2\cdot15. \tag{10.5}$$

This basic sway equation must now be combined with the various beam mechanisms in an attempt to reduce the value of the load factor λ.

Starting with the lower beam, the calculations can be laid out in the usual way:

Sway mechanism (fig. 10.9) $\quad 5760\lambda = 12408; \lambda = 2\cdot15,$ (10.5 *bis*)

L.H. beam (fig. 10.6, eqn (10.3)) $\quad 2720\lambda = 4988$

$$8480\lambda \quad 17396$$

Cancel hinge: $\quad 2494$

$$8480\lambda = 14902; \lambda = 1\cdot76,$$ (10.6)

Centre beam (fig. 10.7, eqn (10.3)) $\quad 2580\lambda = 4988$

$$11060\lambda \quad 19890$$

Cancel hinge: $\quad 2494$

$$11060\lambda = 17396; \lambda = 1\cdot57,$$ (10.7)

R.H. beam (fig. 10.6 reversed) $2720\lambda = 4988$

$$13780\lambda \quad 22384$$

Cancel hinge: $\quad 2494$

$$13780\lambda = 19890; \lambda = 1\cdot443.$$ (10.8)

All the lower beams thus participate in the overall collapse mechanism.

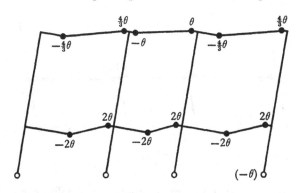

Fig. 10.10

The wind suction on the flat roof relieves the intensity of loading on the roof beam, as will be seen by comparing fig. 10.3 (*a*) and (*b*). The central portion of the roof beam does not form a hinge within its length, but as will be seen below, the two outer spans may be used to reduce slightly the value of λ from that given by equation (10.8). In writing the mechanism equations corresponding to fig. 10.5, the loading terms must be reduced

in the ratio $5/11$ to allow for the smaller loads on the roof; equations (10.1)
are replaced by

$$\begin{aligned}\text{Fig. 10.5}(b): \quad 800 &= 4M_p. \\ \text{Fig. 10.5}(c): \quad 400 &= \tfrac{8}{3}M_p.\end{aligned} \right\} \tag{10.9}$$

The mechanism of fig. 10.5(c) is found to be critical, rather than that of
fig. 10.5(b); thus the calculations may be continued:

$$13780\lambda = 19890; \quad \lambda = 1\cdot443, \quad (10.8\,bis)$$

L.H. roof beam (fig. 10.5(c), $\qquad 400\lambda = \quad 2189$
eqn (10.9))

$$\begin{array}{ll} \overline{14180\lambda} & \overline{22079} \\ \end{array}$$

$$\text{Cancel hinge:} \qquad\qquad 1642$$

$$14180\lambda = 20437; \quad \lambda = 1\cdot441, \quad (10.10)$$

R.H. roof beam (fig. 10.5(c), $\qquad 400\lambda = \quad 2189$
eqn (10.9))

$$\begin{array}{ll} \overline{14580\lambda} & \overline{22626} \\ \end{array}$$

$$\text{Cancel hinge:} \qquad\qquad 1642$$

Final mechanism (fig. 10.10) $\quad 14580\lambda = 20984; \quad \lambda = 1\cdot439. \quad (10.11)$

Thus, subject to the confirmation of a static check, it seems that the
frame of fig. 10.8 will collapse under combined vertical and wind loading
(fig. 10.3(b)) at a load factor of $1\cdot439$; the corresponding mode of collapse
is sketched in fig. 10.10. It will be remembered that the beams were
designed for vertical load only to a nominal load factor of $1\cdot75$; the
nearest real sections available gave an actual load factor of $1\cdot83$, equation
(10.4). Thus this design is just satisfactory in that it provides a load factor
against wind above the nominal value of $1\cdot40$.

There are several ways of making the static check. For example, to
determine the value of the bending moment at joint 22, the *virtual*
mechanism of fig. 10.5(c) can be combined with the (factored) free
bending moments of table 10.1 and with the bending moments of the
collapse state, fig. 10.10:

$$(M_{22})(1)+(-821)(-\tfrac{4}{3})+(821)(\tfrac{4}{3}) = (1\cdot439)(\tfrac{5}{11})(-660)(-\tfrac{4}{3}),$$

that is $\qquad\qquad\qquad M_{22} = -793\,\text{kN m}. \qquad\qquad (10.12)$

(Note that the free bending moments in table 10.1 should be multiplied
by the factor $(\tfrac{5}{11})$ for the wind loading case, to allow for the reduced load
on the roof.)

Similar virtual mechanisms may be used to derive one by one all the necessary values of the bending moments at the critical sections of the beams. Alternatively, the actual bending moments at collapse must be a combination of proper free and reactant bending-moment distributions. Since the reactant line for any beam is straight, and since all beams contain two hinges, it is easy to extend tables such as 10.1. In table 10.3 the reactant bending moments have been fixed at joints 23 and 26 by the condition that the total value, free plus reactant, must equal the full plastic value, 821 kN m. The other reactant values have then been interpolated linearly, and it will be seen that the net values satisfy the yield condition.

Table 10.3

Joint	22	23	24	25	26
Free B.M. ($\times$ 1·439)	0	−432	−576	−432	0
Reactant B.M.	−793	−389	14	417	821
Net B.M.	−793	−**821**	−562	−15	**821**

Table 10.4

Joint	5	6	7	8	
Free B.M. ($\times$ 1·439)	0	−1957	−1612	0	
Reactant B.M.	173	710	979	1247	
Net B.M.	173	−**1247**	−633	**1247**	

Joint	8	9	10	11	12
Free B.M. ($\times$ 1·439)	0	−1392	−1856	−1392	0
Reactant B.M.	−29	290	609	928	1247
Net B.M.	−29	−1102	−**1247**	−464	**1247**

Joint	12	13	14	15	
Free B.M. ($\times$ 1·439)	0	−1612	−1957	0	
Reactant B.M.	173	442	710	1247	
Net B.M.	173	−1170	−**1247**	**1247**	

Similarly, table 10.2 is extended to give table 10.4 for the three spans of the lower beam; again, it will be seen that the full plastic moment of value 1247 kN m is not exceeded. Thus the solution of fig. 10.10 is confirmed as correct for combined vertical and wind loading.

The next step in the preliminary design is to assign provisional sections to the columns. Design of the columns in this example is straightforward, whatever criterion is used; the steps in the calculations will be indicated, but a full analysis will not be given. Starting with the collapse configuration of fig. 10.10, certain bending moments acting at the ends of the members can be determined immediately, fig. 10.11; the bending moments in the beams have been taken from tables 10.3 and 10.4. The frame is collapsing with twelve plastic hinges, and had originally fourteen redundancies; there will thus be three redundancies left at collapse. There must therefore be a single equation connecting the values of the four unknown bending moments marked in fig. 10.11. This equation may

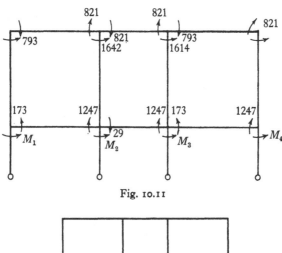

Fig. 10.11

Fig. 10.12

be found by considering the horizontal equilibrium of the frame in the collapse state, that is, by summing the shear forces in the bottom storey:

$$M_1 + M_2 + M_3 + M_4 = (1\cdot439)(3200) = 4605. \qquad (10.13)$$

Equation (10.13) could also have been written immediately by applying the equation of virtual work to the mechanism of fig. 10.12.

The problem now is to determine 'suitable' values of the four unknown bending moments in equation (10.13) in order to establish values for the bending moments in the columns. Assuming that the same section is used throughout the length of each column, and that all four columns are identical, then the ideas of section 5.5 earlier can be used to minimize the largest value of the unknown bending moments in each column. If the beams are imagined to be strengthened artificially, and the loads gradually increased, the final collapse state involves four extra hinges, one in each column as marked in fig. 10.13. Thus, using equation (10.13),

$$M_1 = 1022,$$
$$M_2 = M_3 = M_4 = 1195,$$

(10.14)

and the corresponding values of the bending moments acting on the ends of each column length are marked in fig. 10.14 (which has been constructed from fig. 10.11 merely by entering the values (10.14)).

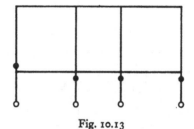

Fig. 10.13

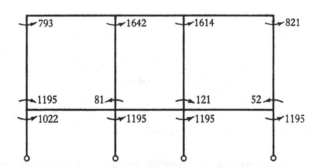

Fig. 10.14

As implied in section 5.5, the distribution of fig. 10.14 is in some sense a 'best case', since any other equilibrium distribution will lead to a bending moment in at least one of the columns whose value exceeds 1195 kN m at the level of the lower beam. As an alternative 'reasonable' set of bending moments for the columns, an elastic analysis will be made of the columns at the point of incipient collapse of the frame as a whole. Now the hinge system of fig. 10.10 effectively isolates each column from its neighbour; due to the formation of hinges in the beams, the *elastic* stiffness of each beam is zero. Thus, referring to fig. 10.11, the out-of-balance moment of 173 kN m at lower-beam level in the left-hand column must be distributed between the upper and lower lengths of the column simply in proportion to their stiffnesses.

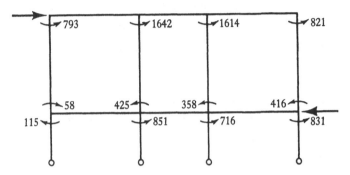

Fig. 10.15

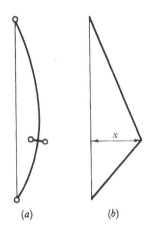

(a)   (b)

Fig. 10.16

With the present dimensions, this stiffness ratio is 2 : 1; the moment of 173 kN m is divided as 58 and 115 kN m, as marked in fig. 10.15. Similarly, the out-of-balance moments for the other columns are divided in the same ratio. The distribution of fig. 10.15 does not, however, satisfy the equilibrium equation (10.13). A familiar situation in the moment-distribution process for sway frames has arisen; the columns have not yet been allowed to bend, since only a straight-line collapse mechanism has been considered. To satisfy equilibrium, an extra geometrical sway must be superimposed on the solution already found. If the lower beam is displaced horizontally relative to the upper beam and the ground, each column will be bent as shown schematically in fig. 10.16(a), and will have a bending-moment diagram of the shape sketched in fig. 10.16(b). If all columns are identical they will have equal stiffnesses, and the value of 'x' in fig. 10.16(b) will be the same for all four columns. Thus the value of x may be found by writing equation (10.13) in the form

$$(x-115)+(x+851)+(x+716)+(x+831) = 4605,$$

that is $$x = 581, \tag{10.15}$$

where the numbers (-115, 851, etc.) occurring in the equation have been taken from fig. 10.15. Thus, superimposing the bending-moment distribution of fig. 10.16(b), for each column, on that of fig. 10.15, the full elastic–plastic solution, fig. 10.17, may be determined.

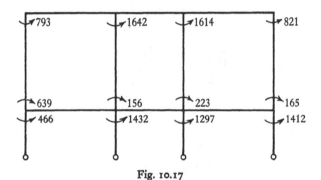

Fig. 10.17

The marked similarities between the values shown in figs. 10.14 and 10.17 imply that the preliminary design of the columns can be undertaken with some confidence, using either distribution as a basis. (The elastic solution from the computer, fig. 10.18 below, while showing some differences from the two present distributions, does not necessitate any change in the column sections.)

The upper length of the left-hand internal column seems critical, with a bending moment of 1642 kN m at one end and a small moment at the other. If a Universal Beam section is used, then only the $36 \times 16\frac{1}{2}$ sections will have a reasonable slenderness ratio over the very long (16 m) unsupported length; the value for the $36 \times 16\frac{1}{2}$ UB 343 is $16000/91\cdot1 = 176$. The corresponding elastic bending stress due to the bending moment of 1642 kN m is $1642 \times 10^6/13\cdot691 \times 10^6 = 120$ N/mm², that is, about one-half of the yield stress. Thus it is clear that a fairly high section modulus is required, and this rules out all but the most heavy of the Universal Column sections, which is nearly twice as heavy as the Beam section noted above. To put the arguments the other way round, the columns must have a large section modulus to carry the high moments due to wind. A suitable choice for this criterion will automatically ensure that the slenderness ratio is reasonable, and that axial stresses are low. Any of the standard criteria for checking the stability of columns will confirm that the $36 \times 16\frac{1}{2}$ UB 343 is suitable for each of the four columns, and it is this section which is marked in fig. 10.8.

This completes the preliminary design of the frame, which must now be examined in the light of other criteria.

10.3   Shakedown analysis of preliminary design

Table 10.5 gives the values of the elastic bending moments at the 45 critical sections of the frame; a separate analysis was made for each of the loading cases noted in section 10.1 and illustrated in fig. 10.2. These results were obtained by using the STRESS program (to which reference has been made) on an IBM 1130 computer; the analysis takes into account deformations due to direct stresses in the members, although these effects are negligibly small in the present example.

The results in the first seven lines of table 10.5 are combined in various ways in the next five lines. The bending moments marked $\mathcal{M}_V$ are those due to all vertical loads acting together, and correspond to the static design case of fig. 10.3(a). Similarly, the bending moments marked $\mathcal{M}_{V+H}$ correspond to the loading of fig. 10.3(b). An immediate check may be made of any set of results in the table by using these elastic bending moments to compute the static collapse equation for any mechanism ϕ:

$$\lambda \Sigma \mathcal{M} \phi = \Sigma M_p |\phi|. \tag{10.16}$$

Thus if the bending moments $\mathcal{M}_{V+H}$ are used with the collapse mechanism of fig. 10.10, equation (10.16) gives

$$\lambda[(-2)(-593)+(2)(1603)+(-2)(-474)+(2)(1595)+(-2)(-545)$$
$$+(2)(1292)+(-\tfrac{4}{3})(-172)+(\tfrac{4}{3})(487)+(-1)(-178)+(1)(466)$$
$$+(-\tfrac{4}{3})(-170)+(\tfrac{4}{3})(474)] = (12)(1247)+(\tfrac{22}{3})(821),$$

that is $14585\lambda = 20984; \quad \lambda = 1\cdot439. \tag{10.17}$

Equation (10.17) is virtually identical with equation (10.11).

Table 10.5

Section	Upper beam, left-hand span					Upper beam, central span			
...	25/22	25/23	26/24	27/25	28/26	29/26	29/27	30/28	31/29
Loading 1	50	−9	−28	−7	53	31	−8	−8	31
2	447	−140	−326	−112	501	288	−112	−112	288
3	16	12	8	5	1	−13	−9	−4	0
3R	−10	−5	0	6	11	0	−4	−9	−13
4	7	1	−4	−10	−16	11	11	11	11
5	−500	−31	198	187	−63	−495	−40	174	149
5R	−36	199	193	−52	−538	149	174	−40	−495
$\mathscr{M}_V$	510	−141	−350	−118	550	317	−122	−122	317
$\mathscr{M}_{V+H}$	10	−172	−152	69	487	−178	−162	52	466
$\mathscr{M}^{max}$	520	203	178	191	566	479	177	177	479
$\mathscr{M}^{min}$	−460	−185	−358	−181	−501	−477	−122	−122	−477
$\mathscr{M}$	—	—	—	—	1067	956	—	—	956

Section	Upper beam, right-hand span					Lower beam, left-hand span			
...	32/29	32/30	33/31	34/32	35/33	5/5	5/6	6/7	7/8
Loading 1	53	−7	−28	−9	50	101	−56	−15	107
2	501	−112	−326	−140	447	18	8	3	−2
3	11	6	0	−5	−10	550	−532	−273	786
3R	1	5	8	12	16	−18	−2	6	14
4	−16	−10	−4	1	7	−28	18	41	64
5	−538	−52	193	199	−36	−691	−29	302	634
5R	−63	187	198	−31	−500	669	19	−306	−632
$\mathscr{M}_V$	550	−118	−350	−141	510	623	−564	−238	969
$\mathscr{M}_{V+H}$	12	−170	−157	58	474	−68	−593	64	1603
$\mathscr{M}^{max}$	566	191	178	203	520	1338	−11	337	1605
$\mathscr{M}^{min}$	−501	−181	−358	−185	−460	−636	−619	−594	−527
$\mathscr{M}$	1067	—	—	—	—	—	—	—	2132

Section	Lower beam, central span					Lower beam, right-hand span			
...	8/8	8/9	9/10	10/11	11/12	12/12	12/13	13/14	14/15
Loading 1	62	−5	−28	−5	62	107	−15	−56	101
2	−7	−7	−7	−7	−7	−2	3	8	18
3	116	71	26	−19	−64	14	6	−2	−18
3R	−64	−19	26	71	116	786	−273	−532	550
4	705	−195	−495	−195	705	64	41	18	−28
5	−775	−385	4	394	783	−632	−306	19	669
5R	783	394	4	−385	−775	634	302	−29	−691
$\mathscr{M}_V$	812	−155	−478	−155	812	969	−238	−564	623
$\mathscr{M}_{V+H}$	37	−540	−474	239	1595	337	−544	−545	1292
$\mathscr{M}^{max}$	1666	460	28	460	1666	1605	337	−11	1338
$\mathscr{M}^{min}$	−784	−611	−530	−611	−784	−527	−594	−619	−636
$\mathscr{M}$	2450	—	—	—	2450	2132	—	—	—

Table 10.5 (cont.)

Section ...	Column A						Column B		
	1/5	15/5	15/16	16/17	17/18	18/22	2/8	19/8	19/26
Loading 1	51	−50	−25	0	25	50	22	−22	22
2	−127	−145	3	151	299	447	−69	−74	213
3	365	−185	−135	−84	−34	16	435	−234	13
3R	−15	3	0	−4	−7	−10	55	−23	11
4	−22	6	6	7	7	7	−417	224	−27
5	−438	253	−175	−494	−552	−500	1128	−281	431
5R	590	−79	172	263	193	−36	−1045	370	−687
$\mathscr{M}_V$	252	−371	−151	70	290	510	26	−129	232
$\mathscr{M}_{V+H}$	−186	−118	−326	−424	−262	10	1154	−410	663
$\mathscr{M}^{max}$	1006	212	156	421	524	520	1640	572	690
$\mathscr{M}^{min}$	−541	−459	−335	−578	−568	−460	−1509	−634	−692
$\bar{\mathscr{M}}$	1547	—	—	—	—	—	3149	—	—

Section ...	Column C			Column D					
	3/12	20/12	20/29	4/15	21/15	21/19	22/20	23/21	24/33
1	−22	22	−22	51	−50	−25	0	25	50
2	69	74	−213	−127	−145	3	151	299	447
3	−55	23	−11	−15	3	0	−4	−7	−10
3R	−435	234	−13	365	−185	−135	−84	−34	16
4	417	−224	27	−22	6	6	7	7	7
5	1045	−370	687	590	−79	172	263	193	−36
5R	−1128	281	−431	−438	253	−175	−494	−552	−500
$\mathscr{M}_V$	−26	129	−232	252	−371	−151	70	290	510
$\mathscr{M}_{V+H}$	1019	−241	455	842	−450	21	333	483	474
$\mathscr{M}^{max}$	1509	634	692	1006	212	156	421	524	520
$\mathscr{M}^{min}$	−1640	−572	−690	−541	−459	−335	−578	−568	−460
$\bar{\mathscr{M}}$	3149	—	—	1547	—	—	—	—	—

The lines denoted $\mathscr{M}^{max}$ and $\mathscr{M}^{min}$ in table 10.5 are merely the computed bending-moments summed to give the greatest and least values at any section; it will be remembered that loading 1 (dead) is always present, loadings 3 and 3R can occur simultaneously, but that loadings 5 and 5R (wind) are alternative. These values of $\mathscr{M}^{max}$ and $\mathscr{M}^{min}$ will be used for a shakedown analysis. The final line of the table, $\bar{\mathscr{M}}$, is simply the value $\mathscr{M}^{max} - \mathscr{M}^{min}$, and only the largest values of $\bar{\mathscr{M}}$ are noted.

Since the loading on the frame is completely symmetrical, in the sense that loading cases 3 and 3R are mirrored about the centre line, and the wind can blow indifferently in either direction, the shakedown analysis is particularly simple. Either alternating plasticity will be critical, or a *symmetrical* incremental collapse mode will govern the analysis.

The question of alternating plasticity will be discussed first. The permitted range $2M_y$ of elastic bending moment at any section may be read from the section tables, using again a yield stress $\sigma_0 = 250\,\mathrm{N/mm^2}$. For the upper-beam, lower-beam and column sections marked in fig. 10.8, the ranges $2M_y$ are 1437, 2182 and 6845 kN m respectively. Using the values $\mathscr{M}$ from table 10.5 the corresponding load factors λ_A which cannot be exceeded if alternating plasticity is to be prevented are

$$\left.\begin{aligned}
\text{Upper beam:} \qquad & \lambda_A = 1437/1067 = 1\cdot35. \\
\text{Lower beam:} \qquad & \lambda_A = 2182/2450 = 0\cdot89. \\
\text{Internal column:} \quad & \lambda_A = 6845/3149 = 2\cdot17.
\end{aligned}\right\} \qquad (10.18)$$

The significance to be attached to load factors against alternating plasticity will be discussed briefly below; however, it would seem that a factor of $0\cdot89$ is probably unacceptable on any basis.

There is also a fairly dramatic drop in the value of the load factor against incremental collapse. The analysis will be done in the form of equation (6.28). Thus, for incremental collapse of the left-hand span of the roof beam (or of the right-hand span) which collapses in the mode of fig. 10.5(b),

$$\lambda_s[(520)(1) + (-358)(-2) + (566)(1)] = (4)(821),$$

that is

$$\lambda_s = 3284/1802 = 1\cdot82. \qquad (10.19)$$

Similarly, the modes of figs. 10.6 and 10.7 for the lower beam give incremental collapse load factors:

Lower beam, left-hand span:

$$\lambda_s[(1338)(1) + (-619)(-2) + (1605)(1)] = (4)(1247),$$

$$\lambda_s = 4988/4181 = 1\cdot19. \qquad (10.20)$$

Lower beam, central span:

$$\lambda_s[(1666)(1) + (-530)(-2) + (1666)(1)] = 4988,$$

$$\lambda_s = 4988/4392 = 1\cdot14. \qquad (10.21)$$

Even without the question of alternating plasticity, it would seem that the section of the lower beam should be increased to raise the load factor against incremental collapse. In fact, however, bearing in mind the arguments of section 6.8 earlier, a factor of $1\cdot14$ might be acceptable, although such a conclusion would have to be closely argued for a particular building on a particular site. Examination of table 10.5 shows that the variations in bending moment for the central span of the lower beam are due largely to the alternative wind loading patterns that have been assumed. The wind loads are themselves high, and there is room for some discussion of the probability of a full gale, from the south-west say, being followed at some later date by an equally heavy gale from the north-east.

However, the lower beam will be strengthened and the frame re-analysed, section 10.4 below. Before doing this, however, the columns may be checked in the light of the elastic solution of table 10.5. The elastic

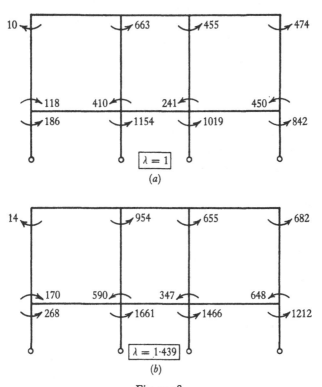

Fig. 10.18

distribution $\mathscr{M}_{V+H}$ is illustrated in fig. 10.18(a); the figure is drawn for a load factor of unity. If all values are multiplied by the collapse load factor of 1·439, fig. 10.18(b), a distribution results which is comparable directly with figs. 10.14 and 10.17. The main differences between fig. 10.18(b) and the other two distributions lie in the values of the bending moments at roof level, and in the windward column. Numerically, the largest bending moment in fig. 10.18(b), 1661 kN m, compares with

1642 kN m in figs. 10.14 and 10.17. The column sections marked in fig. 10.8 are again satisfactory.

In this particular case, therefore, the 'equilibrium' design, fig. 10.14, the elastic–plastic design, fig. 10.17, and the (hypothetical) elastic design, fig. 10.18(*b*), all lead to the same conclusion as regards the stability of the columns.

10.4 Redesign of frame

The preliminary design of fig. 10.8 will be modified in an attempt to improve the performance under variable repeated loading. These modifications are subject to the 'vicious-circle' behaviour noted in section 7.7 with reference to minimum-weight design; shakedown behaviour is essentially an elastic phenomenon, and the section sizes must be known before the elastic response can be computed. However, the elastic response is itself used in assigning member sizes.

In the present case, an attempt will be made to raise the load factors of 1·14 and 1·19 of equations (10.21) and (10.20). The 'obvious' modification is to increase the section modulus of the lower beam, perhaps in the ratio 1·40/1·14; thus the 27 × 10 UB 152 might be replaced by a 30 × 10½ UB 173 section. This would just give the required load factor of 1·40 for the central span of the lower beam, on the assumption that the elastic response of table 10.5 remained the same. Unfortunately, as is usual with trial-and-error methods of elastic design, the changing of a section does *not* leave the elastic response unaltered; moreover, an increase in section very often *increases* the elastic values of bending moment at the points in question. For this reason, the frame of fig. 10.19 will be analysed; the 30 × 10½ UB 173 is used for the outer spans of the lower beam, but the central span uses the heavier section in the same size range.

Results of the complete new elastic analysis are not given here; table 10.6 gives the bending moments in the left-hand and central spans of the lower beam. These two spans continue to be critical, both for alternating plasticity and for incremental collapse. The ranges $2M_y$ for the two 30 × 10½ UB sections ($\sigma_0 = 250$ N/mm²) are 2687 and 3111 kN m; thus, using the values of $\mathcal{M}$ from table 10.6, the load factors against alternating plasticity are

$$\text{Lower beam} \begin{bmatrix} \text{Left-hand span:} & \lambda_A = 2687/2141 = 1\cdot25. \\ \text{Central span:} & \lambda_A = 3111/2674 = 1\cdot16. \end{bmatrix} \quad (10.22)$$

Similarly, the full plastic moments of the two sections are 1546·5 and

1789 kN m. Thus, corresponding to equations (10.20) and (10.21), the two expressions for the load factor against incremental collapse are

Lower beam, left-hand span:

$$\lambda_s[(1350)(1)+(-644)(-2)+(1612)(1)] = 6186,$$

$$\lambda_s = 6186/4250 = 1\cdot45. \qquad (10.23)$$

Lower beam, central span:

$$\lambda_s[(1784)(1)+(-554)(-2)+(1784)(1)] = 7156,$$

$$\lambda_s = 7156/4676 = 1\cdot53. \qquad (10.24)$$

Table 10.6

Section ...	Lower beam, left-hand span				Lower beam, central span				
	5/5	5/6	6/7	7/8	8/8	8/9	9/10	10/11	11/12
Loading 1	99	−57	−15	107	64	−4	−26	−4	64
2	24	11	4	−2	−11	−11	−11	−11	−11
3	534	−542	−280	782	153	96	40	−17	−73
3R	−10	−3	1	5	−73	−17	40	96	153
4	−35	23	52	81	683	−217	−517	−217	683
5	−721	−42	296	637	−870	−430	7	445	884
5R	693	29	−302	−634	884	445	7	−430	−870
$\mathscr{M}_V$	612	−586	−238	973	816	−153	−474	−153	816
$\mathscr{M}_{V+H}$	−109	−610	58	1610	−54	−583	−467	292	1700
$\mathscr{M}^{\max}$	1350	6	338	1612	1784	537	61	537	1784
$\mathscr{M}^{\min}$	−667	−644	−597	−529	−890	−679	−554	−679	−890
$\mathscr{M}$	—	—	—	2141	2674	—	—	—	2674

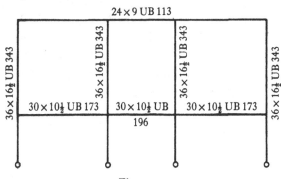

Fig. 10.19

The required load factors against incremental collapse have thus been achieved. No further *static* analysis is required, since it is certain that the

289

frame of fig. 10.19 is 'stronger' than that of fig. 10.8. Further, the bending moments in the columns are little changed.

It remains to discuss the significance of the relatively low values of the load factors against alternating plasticity, equations (10.22). Experimental evidence may be introduced here in addition to the arguments mentioned above about the unlikelihood of the wind acting at full strength in either direction. The analogy with the paper clip, which fractures after a few reversals of bending, is misleading; such fracture occurs only when the plastic strains imposed in each cycle are large. Small plastic strains can apparently be applied repeatedly for many thousands of cycles, and in some cases for hundreds of thousands of cycles.† The alternating hinges formed at a load factor of 1·16 in the present example would suffer plastic deformations whose magnitudes were *contained* by the remaining elastic portions of the frame. Large strains can only occur when a mechanism of collapse is formed.

As has been mentioned, final values of load factors against various conditions for different types of structure can be established only in the light of experience. There is now a wealth of experience with simple steel frames to support the values of static collapse load factor of 1·75 and 1·40 that have been used here. The design of switchhouse in fig. 10.19 would almost certainly be satisfactory, but there is in fact a lack of evidence of the effect of variable repeated loading, both on simple structures, and on more unusual frames such as the switchhouse. For the tall unbraced building even the static loading case requires much more investigation before design load factors can be assigned with confidence. On the other hand, the *braced* multi-storey frame is a straightforward structure; the precision and rationality of plastic methods justify the reduction of load factor from 1·75 to 1·50. Plastic theory can give accurate estimates of the *actual* strength of any of the structures discussed in these two volumes; what is needed in the case of unusual structures is an estimate, based on experience, of the strength that is *required*.

† J. Heyman and B. D. Threlfall, The repeated loading of ductile steel structures, *Proc. Instn civ. Engrs*, **34**, p. 477, 1966.

INDEX

Absolute minimum weight, 202–15
Alternating plasticity, 127–33, 268, 285–6, 288–90
Alternative loading, 196–8, 232–4, 243, 249
Anderson, D., 267
Arches, 69, 73, 75, 78–85
Axial load, 14–16, 18, 46, 63, 69, 243

Baker, Sir John, vii
Bellman, R. E., 167
Biaxial: bending, 21, 45–55; tension, 20
Bounds: lower, 26, 37, 41; on minimum weight, 191–6; upper, 25, 30, 37–8, 41
Box beams, 35
Bracing, 242, 249

Cathedrals, 73, 84
Chan, H. S. Y., 224, 226
Collapse: definition of, 1–6, 101; load, constancy of, 2, 3, 101; mechanisms, alternative, 8
Columns: continuous, 254; design of, 122–3, 249, 254, 267, 278, 287; stability of, 88, 90, 190, 192, 243
Combination of mechanisms, 108, 113–22, 139–42
Compatibility, elastic, 89
Computer analysis, 167, 221, 224, 226, 240, 267
Convexity (of yield surface), 7, 16
Crane girder, 158

Davies, J. M., 158
Deflexions, 88–113, 243, 246; assumed small, 2, 3, 85, 88, 243; of angle cantilever, 50–5, 59
Design plane, 179, 181, 194, 274
Deterioration of stability, 245
Dines, L. L., 222
Dynamic programming, 167–76, 221

Elastic–plastic analysis, 88–113
Equilibrium: condition, 77; state, reasonable, 242

Fatigue, 158
Fenves, S. J., 46, 268
Finlinson, J. C. H., 251, 268
Flow rule, 11, 17, 24–5, 27, 45
Flying buttress, 76

Foulkes, J. D., 183
Foulkes mechanism, 184, 240, 274
Foulkes's theorem, 183–7
Four-storey frame, 108–13, 121
Fowler, P. P., 268
Frame instability, 244
Franciosi, V., 85

'Geometrical' factor of safety, 78, 81, 84
Gillson, I. P., 268
Grillages, 33–42

Heyman, J., 54, 73, 251, 268, 290
Horne, M. R., 158

Incremental collapse, 86, 127–63, 268, 285–6, 288–9
Inglis 'A', 251
Interaction diagram 3, 16, 144, 154–6
Iterative processes of design, 243

Johansen, K. W., 42
Johnson, R. P., 251, 268
Joint Committee on design of multi-storey frames, 252 n
Jones, L. L., 42

Kalker, J. J., 240

Linear: inequalities, 137; programming, 178, 221–6; weight function, 176–83
Livesley, R. K., 224, 225
Load factors, 126, 161–3, 196, 248, 252, 254–6
Loading, alternative, 196–8, 232–4, 243, 249
Logcher, R. D., 268
Lower bounds, 26, 37, 41

Majid, K. I., 267
Masonry, 60, 73–84
Mauch, S. P., 268
Maximum plastic work, 7
Mechanism: condition, 75; over-complete, 115; partial, 88, 92, 115; regular, 88, 92
Middle-third rule, 77
Minimum weight, 115, 167–220, 237–41; absolute, 202–15, bounds on, 191–6
Mises criterion, 12, 20, 21
Moment–curvature relation, 89, 129, 132